The Last of the Welsh Coal Trains

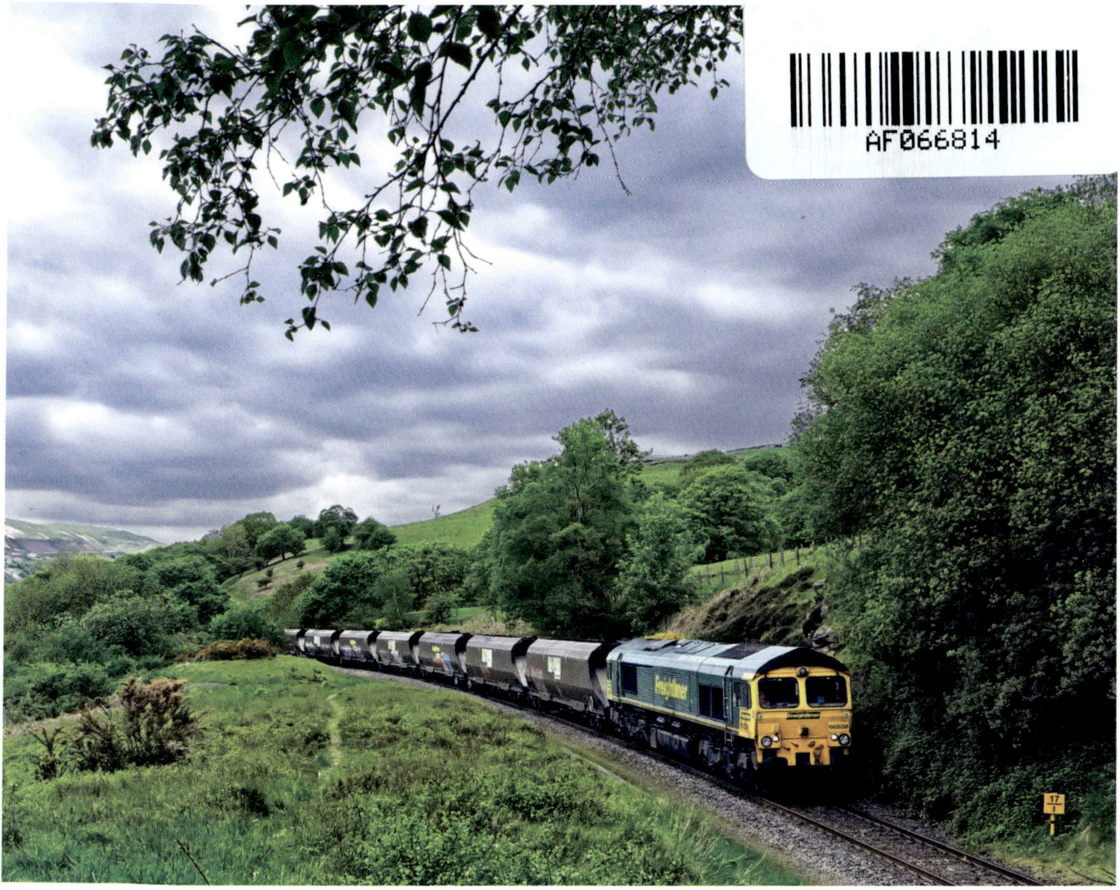

CHRIS DAVIES

KEY Books

THE RAILWAYS AND INDUSTRY SERIES, VOLUME 2

Title page image: A view from the lineside, looking north along the Taff Bargoed Valley on 18 May 2018. Class 66/5 No. 66538 approaches the village of Bedlinog on the 14:00 (6C93) to Port Talbot Grange with around 1,500 tonnes of coal for Tata steel works. The coal will be converted into coke for smelting iron ore.

Contents page image: The final destination! On 11 November 2016, Class 66/5 No. 66519 makes its way past the vast stockpile of coal at Aberthaw 'B' power station to discharge its load of coal brought in from Avonmouth on the 6B68.

Published by Key Books
An imprint of Key Publishing Ltd
PO Box 100
Stamford
Lincs PE19 1XQ

www.keypublishing.com

The right of Chris Davies to be identified as the author of this book has been asserted in accordance with the Copyright, Designs and Patents Act 1988 Sections 77 and 78.

Copyright © Chris Davies, 2020

ISBN 978 1 913295 73 8

All rights reserved. Reproduction in whole or in part in any form whatsoever or by any means is strictly prohibited without the prior permission of the Publisher.

Typeset by Aura Technology and Software Services, India.

Contents

Introduction 4
Chapter 1 The Rise and Fall of Welsh Coal 7
                When Coal Was King 7
                Decline of the Industry..8
                Going Greener 9
                Death of an Industry..9
                The Demise of Coal Rail Freight 11
                Outlook for the Future 12
Chapter 2 South Wales Coal Flows – 2013 to 2020...15
Chapter 3 Cwmbargoed 17
Chapter 4 Tower 51
Chapter 5 Onllwyn 64
Chapter 6 Gwaun-Cae-Gurwen 94
Chapter 7 Aberthaw Power Station 100
                Rail Traffic 101

Introduction

I recall seeing MGR coal trains hauled by Class 37 locomotives heading for Aberthaw power station passing through Llanishen, Cardiff Queen Street and Cardiff Central stations back in the 1980s when, having just graduated from university, I lived in Lisvane in Cardiff with my parents, but I just took them for granted. Later, when I was on leave in the mid-1990s, I visited Barry station which made a big impression on me. It had an old-fashioned, calm feel to it, replete with semaphore signals, including an impressive three-arm gantry semaphore at one end of platform 2 and a signal box at the other. Add to that the Class 37 locomotives stabled at the Barry wagon repair shop on the other side of the platform and I couldn't get enough of it! Then, the innermost arm of the GWR lower quadrant gantry signal dropped down to show a clear line to Aberthaw, so I knew a train was coming, and the peace was shattered by the thundering roar of two Class 37s passing through the station, hauling a coal train to Aberthaw. I had no idea where the train had come from, other than somewhere in the Valleys, but wish now I had had the common sense to return with a camera, but I never did until many years later. Hopefully, I have made up for the deficit, even though the coal trains are no longer hauled by the stalwart Class 37s.

I have been interested in railways since I was a little boy and am old enough to just about remember the last of the steam engines in action, mainly shunting at Barry in South Wales. Whilst I have long been a keen photographer, it is only this past few years, now that I am semi-retired, having spent many years overseas with my career as a geologist, that I have started photographing railways.

I thought a good place to start would be Barry. Alas, the 37s had long gone, but coal trains were still running on a regular basis to Aberthaw. This was in 2013, when one could still see seven or more trains a day bringing in coal from Onllwyn, Cwmbargoed, Tower, Avonmouth and even from as far away as Scotland. Porthkerry Viaduct was a real favourite of mine (and still is), but then I started going further afield, making visits to Cwmbargoed, Tower, Onllwyn, Pontypridd and to Pantyffynnon to record the trains and the last remnants of coal mining in South Wales.

I soon realised the days of coal trains on the rail network were numbered as green energy policies, which included the phase out of all the UK's coal-fired power stations by 2025, were implemented to reduce emissions as a consequence of climate change. I think it's fair to say that just a few years ago, railway photographers took coal trains for granted as they were so common on the network. How that has changed!

With this in mind, I set about recording the trains in as many photogenic locations as I could find in South Wales, over a period of around six years from late 2013 to early 2020. For over a century, coal has been mined in South Wales and then despatched by rail for domestic consumption, and it is one of the last places in the UK where this still takes place. This spurred me on to keep taking shots before it's all confined to the history books.

With the scarcity of the coal freight (one or two trains a day from the mine sites) and our inclement weather, it has been a challenge, but a task that has kept me motivated. Some of the locations featured have been visited many times to achieve the ideal lighting conditions. It's not just the trains that interest me, I am also keen on capturing the environment the train is in to put the shots in context geographically, to hopefully imbue a Welsh feel.

One of my favourite spots is the remote outpost of Cwmbargoed, near Merthyr Tydfil. With its bleak, open, desolate countryside, there are some excellent photographic opportunities. It has a charm all of its own and it's enthralling to see how the environment changes over the seasons. One of my favourite times to visit is the autumn when the russet-brown bracken dominates the Taff Bargoed Valley. Equally, winter views can also be special when the snow-clad hills are punctuated by a 'red train' departing or going to the washery, though driving in such conditions is somewhat nail biting! The steep climb the line takes along the valley can also make for some really impressive photography in the unspoiled landscape just south of the washery.

The ever-changing inclement weather conditions, however, can pose challenges for the ardent railway photographer. Cwmbargoed seems to have its own microclimate where one moment the landscape can be bathed in sunlight, and the next it will disappear completely when moving clouds obscure the sun, shadowing the landscape. It can be frustrating, but with a little patience, extremely rewarding too. I often find the best time to go out is during periods of showers as they can lead to some great lighting conditions and stunning skyscapes. One bonus is that the open countryside provides numerous locations to photograph the trains, a rare scenario on the network nowadays as prolific vegetation frequently reduces photographic opportunities. Shots of the same train travelling up or down the valley can be gained from the same vantage point, but fitness is required to get to some of these!

Onllwyn in the Dulais Valley is in some ways like Cwmbargoed. It too has a strong sense of openness and remoteness, and again affords several views of a single train from one vantage point, particularly when a train is loading up or discharging coal at the wash plant. It is an ideal location, given the paucity of trains.

The Vale of Glamorgan line, which leads to Aberthaw power station and is also used for the trains to Tata steel works in Port Talbot, is also another wonderful location to visit. Porthkerry Viaduct, comprising some 16 arches, between 45 and 50 feet (15 metres) in width, and rising to a height of 110 feet (33 metres), is a spectacular feat of engineering which I never tire of photographing. Again, it is fascinating to capture the changing mood over the seasons, an easy feat for me as I live nearby. I particularly like the early morning winter sunrises illuminating the arches, or capturing the last light catching the arches on a balmy summer evening. I am glad I made the most of the trains crossing the viaduct when they were running on a regular basis a few years ago. It's difficult to replicate such scenes today due to substantially less traffic running as a result of the closure of Aberthaw power station.

The section of coast in the Vale of Glamorgan at Fontygary, west of Barry, is another special place, especially for sunset shots. It was voted number 18 of the top 25 best railway locations to photograph in the UK by the *Rail Express* magazine in May 2014 (edition No. 216).

Apart from one image taken at Earles Sidings in Derbyshire, the images featured in the book are all from coal flows within South Wales. I have also included shots of trains bringing in coal from outside the principality that was destined for Aberthaw but have not included the coal flows along the Marches line to Fiddler's Ferry power station.

The images were taken using a Fujifilm compact SLR camera, spanning several models, including the X-E1, XT-10, XT-20 and XT-3. They were shot as RAW files and processed using Photoshop. For convenience, the images are featured from east to west, starting with Cwmbargoed, the most easterly location, through to the most westerly location at Gwaun-Cae-Gurwen in the Amman Valley. The exception is Aberthaw power station, which features in the final chapter.

I have enjoyed putting this book together and it has entailed extensive research, mostly based on information gleaned in the public domain. Every effort has been made to keep it as accurate as possible but no doubt there are some mistakes, for which I apologise. I would be more than happy to hear from anyone with any comments.

I should like to thank Mark Rowlinson of Freightmaster Publishing, who was kind enough to let me use his coal flow image. The Freightmaster Publishing forum has also proved an invaluable source for much of the information used in the book, especially the posts compiled by Rowland Pittard, who has an encyclopaedic knowledge of freight workings in South Wales. Martin Buck's *Loco Review*, editions 2016 and 2017, have also been very helpful. I am also indebted to Adrian Kenny, who wrote an excellent article for *Railways Illustrated* in 2017, titled 'King Coal No More, a South Wales coal review during the period 2016–2017' (June 2017, Issue 172 & July 2017, Issue 173). The article was a huge help to me, particularly regarding some of the technical aspects of the train operations. Michael Rhodes' book, *Freight Trains of British Rail*, published in 1982, provided very useful historical information about coal traffic flows to Aberthaw. There is a wealth of information on coal on the internet which I also found very helpful. In particular, published statistics available from the UK government on freight rail usage, coal consumption in power stations, and coal production from mining in the UK.

Finally, I should like to thank Key Publishing who approached me to compile the book last November, and my wife, Claire, who was kind enough to read through the manuscript and make amendments where necessary. A tall feat for a non-railway person!

Chris Davies
12 March 2020

On a snowy 12 December 2017, 66158 heads through the village of Bedlinog as it climbs up the Taff Bargoed Valley towards Cwmbargoed, working the 4C94 from Margam.

Chapter 1

The Rise and Fall of Welsh Coal

When Coal Was King

Coal is a sedimentary rock composed predominantly of carbon that is readily combustible and has been a source of heat generation for centuries. It is formed from plant remains that have been compacted, hardened, chemically altered, and metamorphosed by heat and pressure over millions of years. There are three main types of coal, classified by their carbon contents: anthracite, which is a high-grade coal with a high percentage of fixed carbon and low percentage of volatile matter; bituminous coal, commonly referred to as steam coal in the UK, which has a high heating value and is most commonly used for electricity generation; and, lignite, the lowest-grade coal with the least amount of carbon.

From west to east, the South Wales Coal Measures stretch for nearly 145 kilometres from Pembrokeshire, through Carmarthenshire, Swansea, Neath, Port Talbot, Bridgend, Rhondda Cynon Taff, Merthyr Tydfil, Caerphilly, Blaenau Gwent and Torfaen. Geologically, the coal measures occur in an elongated basin (syncline) rimmed by Carboniferous limestone and millstone grit, with upper Carboniferous coal seams of varying thickness associated with sandstone units (Penant sandstone) and mudstones occupying the basin. The coal increases in grade (or rank) from east to west. Bituminous coal is dominant in the east, higher-grade bituminous (steam coal) is found in the central coalfields and high-grade anthracite coal generally occurs in the west.

The South Wales coalfield (sometimes referred to as 'the Valleys') describes the area covered by numerous individual collieries along the valleys in South Wales. Coal mining contributed significantly to the Welsh economy from the mid-nineteenth to the late twentieth century and was also key to the industrial revolution. Deeply incised valleys are a distinctive feature of the Valleys. They are drained by a series of south and south-eastward flowing rivers which have cut through the coal measures, thereby allowing access for the exploitation of the coal, via drifts, adits and vertical shafts. Numerous pits (collieries) were constructed in the Valleys and new railways were built to take the coal to the coast for export from the new ports of Barry and Cardiff. Employment grew, immigration soared and entrepreneurs responsible for developing the coalfields made huge profits.

The early twentieth century was the period when 'coal was king'. From 1913 to 1926, the South Wales coalfield was the largest producer in the UK and the world's largest coal-exporting coalfield. Welsh steam coal from the Rhondda Valley was highly prized for use in steam locomotives and for use in the Royal Navy's battle ships. By 1891, 30 million tonnes of coal had been mined in South Wales and by 1913, when more than a quarter of a million miners were employed, this increased to 57 million tonnes, one fifth of UK coal production. The same year also resulted in record exports from Barry Docks, which became the busiest coal exporting port in the world, when 11.05 million tons (11.23 million tonnes) of coal left the port, overtaking Cardiff Docks.

At its peak in 1920, the coal industry in the UK employed 1.2 million workers, more than one in 20 of the UK work force.

Decline of the Industry

The decline of the British coal industry started after the First World War and accelerated after the Second World War. The South Wales coalfield saw a decrease in production as a result of the depression and declining use for coal. From 1921 to 1936, 241 collieries closed in the Valleys with the loss of 140,000 jobs. However, the industry enjoyed a brief reprieve during the Second World War due to rearmament and the dramatic increase in demand for coal.

The government took control of the mines in 1947 under a Labour government when the coal mines were nationalised and run by the National Coal Board (NCB). Under nationalisation in the UK, pits were modernised, working conditions improved and wages increased. Coal continued to be in demand and production output from South Wales was 24 million tonnes. During the 1950s, the 1956 Clean Air Act coupled with the increase in the use of alternative energy fuels, such as nuclear and gas, resulted in the closure of 25 pits between 1953 and 1959. A further 35 collieries closed between 1960 and 1965. The programme of pit closures continued into the 1960s, with a focus on keeping the larger pits in production whilst closing the smaller, uneconomic pits.

In the early 1970s, a strike over wages collapsed due to a lack of national union recognition. However, this changed in 1972 when the National Union of Mine Workers (NUM) declared an official strike which crippled production and coal transport throughout the UK. A court enquiry recommended wage increases but this never materialised and continued pit closures led to further job losses in South Wales. As a challenge to the government, miners worked to rule, resulting in significant loss of production. This culminated in coal rationing and the Conservative government under Edward Heath declared a state of emergency followed by a three-day working week.

A General Election was called in 1974 and a Labour government was elected. The situation improved under the new Labour government for a short while as many collieries were upgraded, and new coal seams were mined. Only a handful of collieries closed during this time in South Wales as the Labour government was keen to expand the industry. However, conditions were still very challenging, and the government proposed a 'Social Contract' to contain wages and support its policies. Nonetheless, the miners of South Wales voted against this policy in favour of more strikes.

During the 1980s, under a new Conservative government with Margaret Thatcher as Prime Minister, support for the coal industry declined significantly. The Conservative government accepted a Commission report to close 27 of the remaining 33 collieries in South Wales. Miners voted for an immediate strike but due to low coal stocks the government did not immediately implement the closures. Instead, determined to avoid a repeat of the 1970s, it built up coal stocks for power stations nationwide in anticipation of further industrial action and focussed on nuclear energy. On 6 March 1984, the NCB announced that it intended to close 20 pits nationwide with the loss of 20,000 jobs. Consequently, on 12 March 1984, Arthur Scargill, president of the NUM, called a national strike in response to the proposed pit closures.

The 1984–85 strike was one of most bitter industrial disputes the UK had ever seen, as more than half the country's 187,000 miners went on strike. Not all the mines ceased production though. In Nottinghamshire, most miners chose to carry on working, leaving whole towns and villages divided between strikers and those who returned to work. Welfare benefits were removed from striking miners and bonuses were paid to miners who went back to work. There were ugly clashes between the police and the miners and violence reached unprecedented levels.

The strike was unsuccessful, ending on 3 March 1985, when most strikers had returned to work. It was a defining moment in British industrial relations as the NUM's defeat greatly weakened the trade union movement. It was, however, a significant victory for Margaret Thatcher and her Conservative government.

Pit closures nationwide accelerated and by 2009 only six working pits operated in the UK. In South Wales, the last deep pits to close were Taff Merthyr (1992) and Tower Colliery (2008). When the last three deep mines in the UK closed in 2015 (Hatfield, Kellingley and Thoresby), coal mining was confined to just a few pockets of opencast pits in South Wales and Ayrshire.

One of the legacies of coal mining in South Wales today is that a substantial rail network has survived, enjoyed today by thousands of commuters to Cardiff and other valley destinations. Indeed, Network Rail's Route Study for Wales identified that demand for rail has grown from 20 million passenger journeys in 2004/2005 to 30 million today.

Going Greener

To tackle climate change, the UK's goal is to become substantially greener by eliminating the burning of coal for energy in order to reduce CO_2 emissions. Under the 2008 Climate Act, the UK set legally binding targets to reduce CO_2 emissions by at least 80 per cent by 2050. This has since been superseded by the Net Zero Emissions law, which requires the UK to bring all greenhouse gas emissions to net zero by 2050. The UK is now taking the lead on climate change.

As the reliance on coal decreases, the last decade has seen a shift to renewable energy, made up of hydropower, waste, solar, biomass and the largest growing area, wind. In 2017, renewables supplied more than ten per cent of the UK's total energy. Scotland is leading the way with 74 per cent of its gross electricity derived from renewable sources in 2018, according to *Power Technology*.

By 2016, coal-fired power stations supplied just nine per cent of the power generated in the UK, which fell to five per cent in 2018, and by 2019, to just one per cent of all electricity generated. Most of the power now being produced is from gas, oil, nuclear and renewable energy. In fact, in May 2019, the UK went a whole week without coal-generated power, the first time since the Industrial Revolution!

Death of an Industry

In November 2015, it was announced by the UK government that all the nation's coal-fired power stations would close by 2025. Recent closures have included Ironbridge in 2015, followed by Longannet, Ferrybridge and Rugeley in 2016. In 2018, Eggborough closed but was granted consent to convert to a gas-fired power station. Lynmouth power station converted to biomass and Uskmouth is being converted to a waste-energy plant. Aberthaw and Cottam both shut down operations in 2019, whilst Fiddler's Ferry will close in 2020. At the time of writing, Drax, West Burton and Ratcliffe are still operating.

It is interesting to look at the decline of coal used in power stations in the UK over the past 30 years. In 1990, coal-power generation was 84 million tonnes, which halved to 41.5 million tonnes by 2010. In 2015, it fell to a record low of just 29 million tonnes, 22 per cent lower than coal consumed in 2014. This was due to the introduction in 2015 of the UK's top-up carbon tax, a measure implemented to encourage power stations to use greener fuels. It meant a doubling of the price per tonne of CO_2 emissions from £9.54 to £18.08 per tonne. Factoring in the EU's emissions trading system, this raised the cost of a tonne of CO_2 produced by British power stations to £23. Essentially, this meant that electricity generation was no longer competitive and sustainable, forcing power stations to close.

In response to the rapid decline in demand for coal, surface mine production fell to just 2.6 million tonnes in 2018, an 87 per cent decrease from production levels in 1990. As one analyst recently put it: 'coal's fall from grace in the UK is without parallel and the speed at which the industry has declined is unprecedented.'

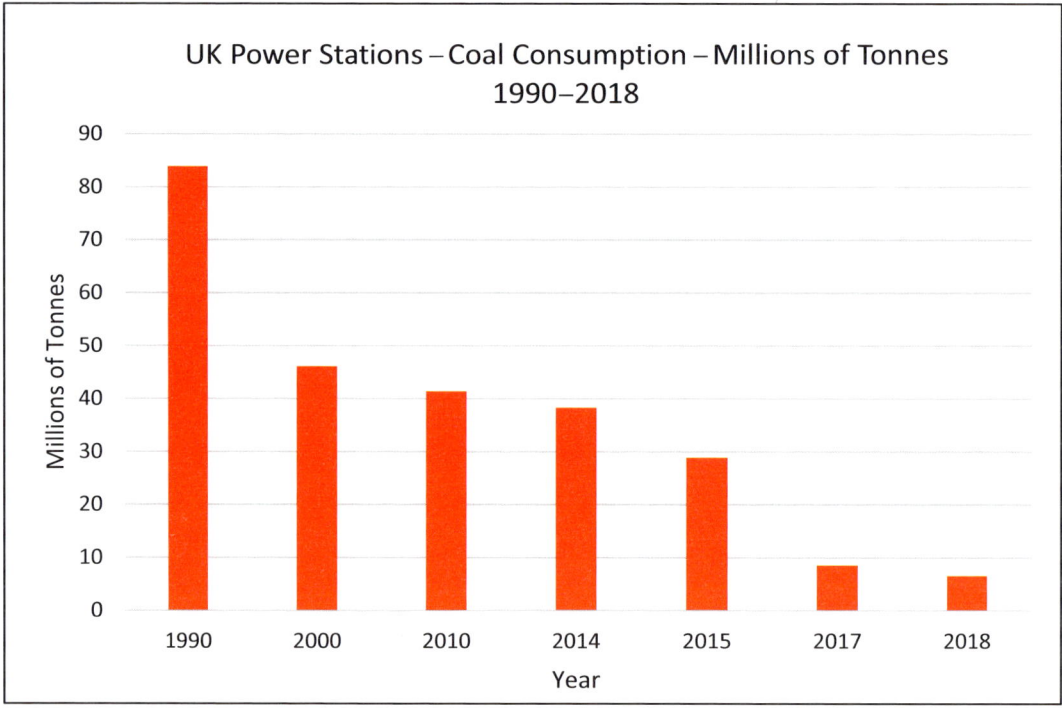

Source: https://assets.publishing.service.gov.uk/government/uploads/system/uploads/attachment_data/file/857027/UK_Energy_in_Brief_

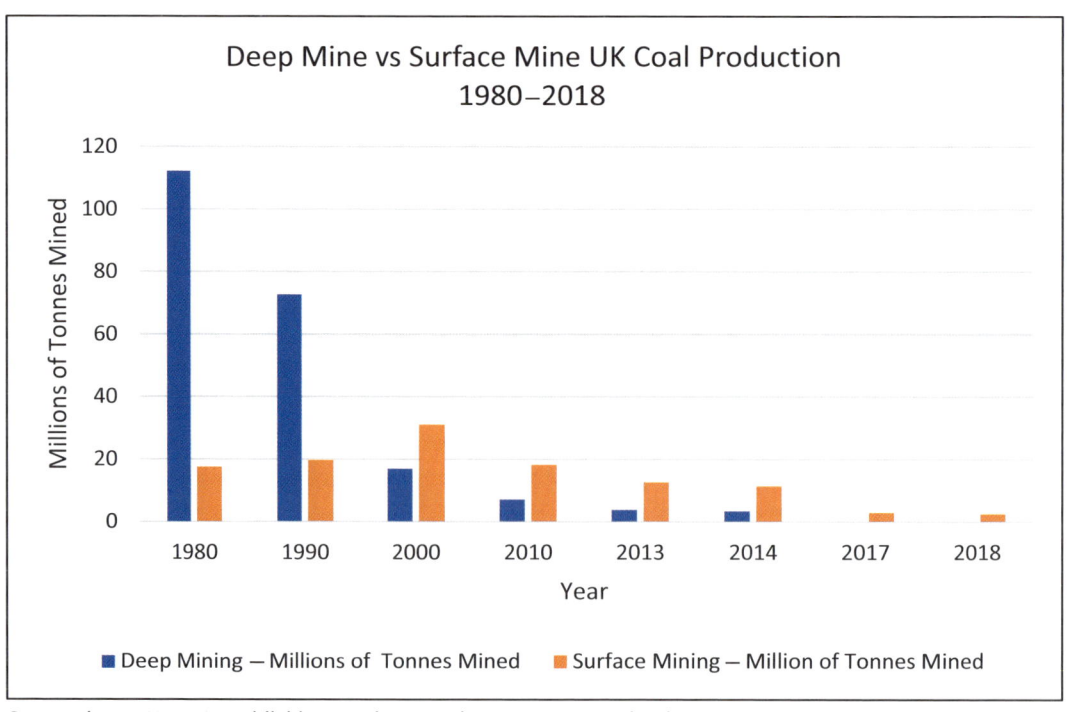

Source: https://assets.publishing.service.gov.uk/government/uploads/system/uploads/attachment_data/file/857027/UK_Energy_in_Brief_

The Demise of Coal Rail Freight

For decades, coal was the cornerstone of the UK rail freight business. Railways were the only means to transport heavy goods in bulk and roads were not an option for many years. In the mid-1930s, around 65 per cent of all rail freight tonnage was coal or coke and this remained at 60 per cent, or about 154 million tonnes, for the next 20 years.

Despite the decline in the latter years of deep coal mining in the UK, coal was still a vital bulk commodity energy source for the power stations. One of the few innovations of the Beeching Act was the introduction of the Merry-Go-Round (MGR) trains in the 1960s, conveying coal from pithead and port to the power stations. The trains travelled in a circuit between the mines and power stations, and were timetabled in between passenger services. As mines closed, more reliance was placed on imported coal from ports such as Grimsby and Immingham to supply the power stations. Two axle air-braked 32-ton (29 tonnes) capacity HAA coal hoppers, constructed between 1964 and 1982, were used on the trains and proved to be one of the most successful wagons ever built; they were finally phased out in 2010.

Coal imports reached a new record level of 50.5 million tonnes in 2006 and remained steady until 2013. The coal freight was a welcome boost for the newly privatised freight operating companies from the mid-1990s to the mid-2000s. Between 2005 and 2015, rail freight operating companies were moving an average of 45 million tonnes of coal a year, representing around 7.2 billion tonne kilometres. Indeed, between 2013 and 2014, coal still accounted for 36 per cent of all rail freight as 51.5 million tonnes of coal was moved, equating to 8.1 billion net tonne kilometres.

Freight operating companies invested in new rolling stock to cope with the demand in the coal traffic. EWS (later DB Cargo) had the largest of the new purpose-built fleets comprising 1,145 HTA wagons between 2001 and 2004, with a 76-tonne capacity, while Freightliner initially purchased 446 HHA wagons (73.6-tonne capacity) and later acquired 220 HXA wagons (74.5-tonne capacity). Soon after, GBRf and Jarvis Fastline also acquired their own fleets of hopper wagons.

The post-privatisation coal rail freight boom came to a swift and abrupt end though when the amount of coal carried by Britain's railways plummeted due to the doubling of the top-up carbon tax in

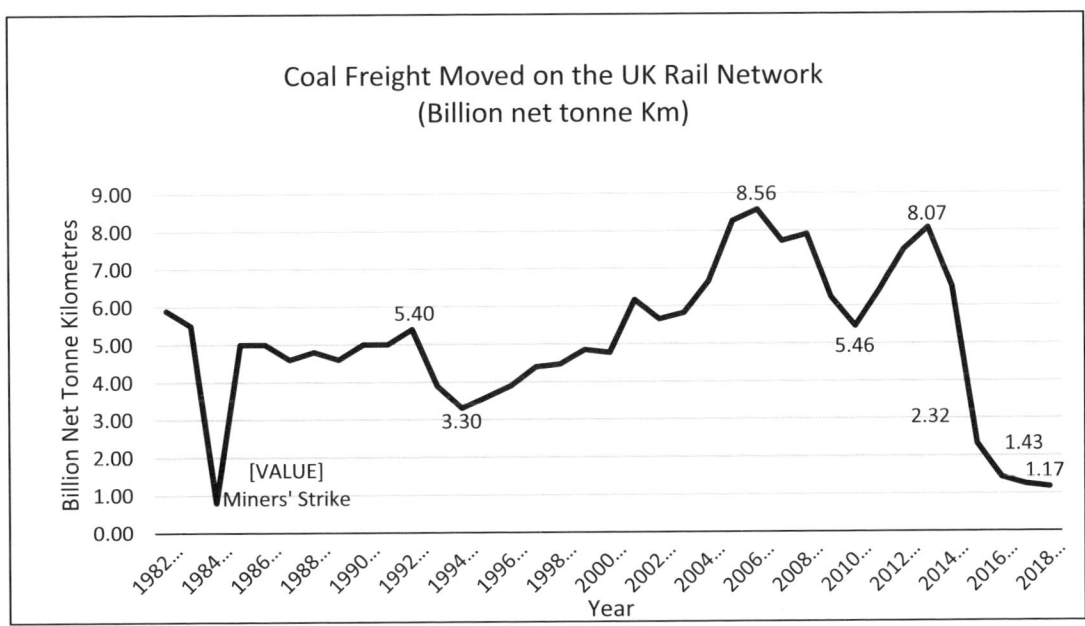

Source : https://dataportal.orr.gov.uk/media/1403/freight-rail-usage-2016-17-quarter-4.pdf

April 2015, leading to the closure of many power stations. From 2015 to 2016, the total amount of coal moved on the network fell 55 per cent to a 30-year low of 19.8 million tonnes, and volumes plummeted 64 per cent to 2.3 billion net tonne kilometres. The dramatic fall in coal freight was described as 'falling off a cliff' by one Rail Freight Group Director. Between 2018 and 2019, coal freight was just 10.5 million tonnes, an 80 per cent decrease from the period 2013–14.

Previously, the largest commodity group on the network, coal is now substantially behind domestic intermodal and construction freight. Many of the HTA and HHA wagons have become surplus to requirements, though some have been converted to carry aggregates. It remains to be seen what will happen to the remaining coal traffic on the network over the next decade or so.

Outlook for the Future

Coal has been at the heart of the UK's economic development for the past two centuries but could soon largely be confined to the history books. Due to the increased awareness of climate change and the environmental damage caused by coal and other fossil fuels, there has been a concerted effort to stop using fossil fuels for energy generation. This includes the government's mandated phase-out of all coal-fired power stations by 2025, one of the Conservative party's flagship green policies.

However, as coal-fired power stations like Aberthaw close, it's worth noting that Britain is still reliant on importing coal-generated electricity from the EU, particularly from the Netherlands and Germany, which means offshoring the greenhouse gas emissions into Europe. In recent years, Germany has opened new coal-fired power stations and does not plan to phase them out until 2038. Offshoring greenhouse gas emissions is an issue the UK government faces and must address.

While decarbonising energy is crucial to obviate climate change, the UK's CO_2 emissions and coal production statistics need be put into context. The UK was ranked 33rd amongst the coal producing countries worldwide in 2018, way behind the top ten largest producers – China, India, the USA, the EU, Australia, Indonesia, Russia, South Africa, Kazakhstan and Colombia. In terms of CO_2 emissions, the UK was reported to produce one per cent of world emissions in 2017, a fraction of China's at 29.34 per cent.

Coal is still vital for many industrial applications and, no matter how desirable it might be, it cannot be dispensed with altogether at the present time. If all the UK's power stations were to close immediately, the UK would still need between 5 and 6 million tonnes per annum to supply a range of essential industries, especially steel and cement manufacturing.

The steel industry is a major consumer of metallurgical or coking coal, a vital ingredient for the smelting of iron ore in blast furnaces. Coal is converted to coke by driving off impurities to leave almost pure carbon. This is achieved by heating the coal to around $1000–1100^0C$ in the absence of oxygen. The industry depends heavily on coal to fuel this energy-intensive processes; for every 1¼ tonnes of steel produced, one tonne of coal is required. The construction of the new HS2 line illustrates this well. An estimated 2 million tonnes of steel will be required to build the railway, which in turn will require 1.6 million tonnes of coal.

The UK and European steel-making industry imported some 52 million tonnes of seaborne metallurgical coal in 2019 from the United States of America, Australia, Russia and Colombia. While an increasing amount of steel is being recycled, there is currently no technology to make steel on a large scale without using coal. Electric-fired blast furnaces can only be used to recycle steel.

Coal is also a vital energy source in cement production with around 200kg required to produce one tonne of cement, and about 300–400kg of cement is needed to produce one cubic metre of concrete.

It is also used in agricultural products, including fertiliser, pesticides and fungicides and has potential application in the production of carbon fibre.

The UK currently imports 80 per cent of the coal that it needs for its industry, but this comes at a price. Imported coal shipped from tens of thousands of miles away is a large source of greenhouse gas emissions. Research by the New Centre for Policy Studies shows that on a per tonne of coal basis, importing coal to industrial customers from Siberia (via St Petersburg port) results in 569 per cent more in CO_2 equivalent emissions than transporting coal mined in the UK. Statistics put the USA at 235 per cent and Colombia at 211 per cent. Worse still, is coal imported from Australia, which at 625 per cent shows the huge carbon footprint associated with bringing coal in from so far. More than 600,000 tonnes of Australian coal was imported to the UK in 2019.

In this connection, it is worth examining what is happening in Wales. Port Talbot steel works relies heavily on coal imports from overseas, which were reported at 1.85 million tonnes in 2018. Locally sourced coal from South Wales is also an important supply. In recent years, Cwmbargoed, near Merthyr Tydfil, has made a significant contribution, despatching one to two trains per day (Mondays–Saturdays) to Tata's plant. Tata also receives indigenous coal from Northumberland. Indeed, such is the importance of locally sourced coal to meet demand, Tata even investigated the feasibility of mining coal from drifts just 5km from the plant but did not proceed with the venture.

The Welsh government has declared a climate emergency and is committed to achieving a carbon neutral public sector by 2030. To meet this strategy, new coal mining applications in Wales are set to be rejected as a matter of policy under new recommended planning rules. The draft policy states: 'Proposals for opencast, deep-mine development, or colliery spoil disposal, should not be permitted.' These are recommended changes to planning policy set by the Welsh National Assembly's Petitions Committee in December 2018. Planning Policy Wales governs what councils can allow through planning permission.

The proposed policy also states: 'Should, in wholly exceptional circumstances, proposals be put forward, they would clearly need to demonstrate why they are needed in the context of climate change emissions reductions targets and for reasons of national energy security.'

Planning permission to mine coal at Cwmbargoed expires in 2022, so it remains to be seen if an extension will be granted for the coaling operation to continue after that. In the long term, the operating company, MSW, hopes to mine coal from the nearby Nant Llesg deposit, near Fochriw, west of Rhymney, which has an additional 6 million tonnes of reserves. However, an application lodged by MSW seeking planning permission to extend the current operation after 2022 was rejected by Caerphilly Council. A subsequent planning appeal was also unsuccessful and there has been huge opposition to the proposed mine by local residents.

Coal imports result in far higher greenhouse gas emissions than the mining of British coal. To offset the emissions, one solution would be to maintain the supply of British coal to meet, as far as possible, the nation's industrial needs. A logical approach would be to continue mining sustainably, whilst transitioning in the long term towards more renewable energy to supply our coal-intensive essential industries.

Welsh coal constitutes an important supply for Tata's coal requirements at Port Talbot. Cwmbargoed is just 37 kilometres (23 miles) away from Port Talbot as the crow flies and could continue to supply Tata for the foreseeable feature. If this supply is supplemented with coal from other mines in the UK, such as the new Woodhouse colliery in Cumbria, it could even lead to a reduction of imports at Port Talbot. Either way, a substantial amount of coal is still required for the UK industrial market.

The future of coal mining in Wales now hangs in the balance. The proposed changes under Planning Policy Wales do, however, acknowledge that under exceptional circumstances new mining projects might be allowed to proceed if warranted. After decades of coal mining in the principality, it remains to be seen if the industry will soon come to an end, as we move away from fossil fuels towards more eco-friendly forms of energy.

In contrast, elsewhere in the UK, a different scenario is being played out. In Cumbria, the UK's first deep coal mine for more than 30 years will be developed by West Cumbria Mining (WCM), who are developing an underground metallurgical coal mine, off the coast near Whitehaven, to supply the UK and European steel-making industry. Formal planning approval for the development of the mine (the Woodhouse Colliery) was granted in 2019 by Cumbria County Council. WCM has committed to move all the coal by rail, either to steelworks in the UK, or to the Redcar Bulk Terminal for export to Europe. When the mine is fully operational, it will produce 2.4 million tonnes of coal a year. Up to six trains a day will be required to transport the coal out of the mine. The traffic is expected to be hauled by Freightliner.

On 7 September 2017, 66004 is seen loading up with coal at the Onllwyn wash plant for the onward journey to Immingham (6E09).

Chapter 2
South Wales Coal Flows – 2013 to 2020

There were steady coal flows in South Wales until the decision to stop burning Welsh coal at Aberthaw in early 2017. Coal was mainly sourced from the last of the operating open-pit mines at Cwmbargoed and Tower and from pits supplying the Onllwyn distribution centre (Nant Helen and East Pit). Imported coal from Avonmouth and Newport Docks to supplement the Welsh coal burned at Aberthaw were also important flows.

Coal traffic diminished considerably when Aberthaw 'B' power station stopped burning Welsh coal in 2017 and then closed in 2019. However, there are still important flows from Cwmbargoed to the Tata steel works in Port Talbot and to British Steel's Scunthorpe works. In addition, coal is despatched to Hope cement works in Derbyshire.

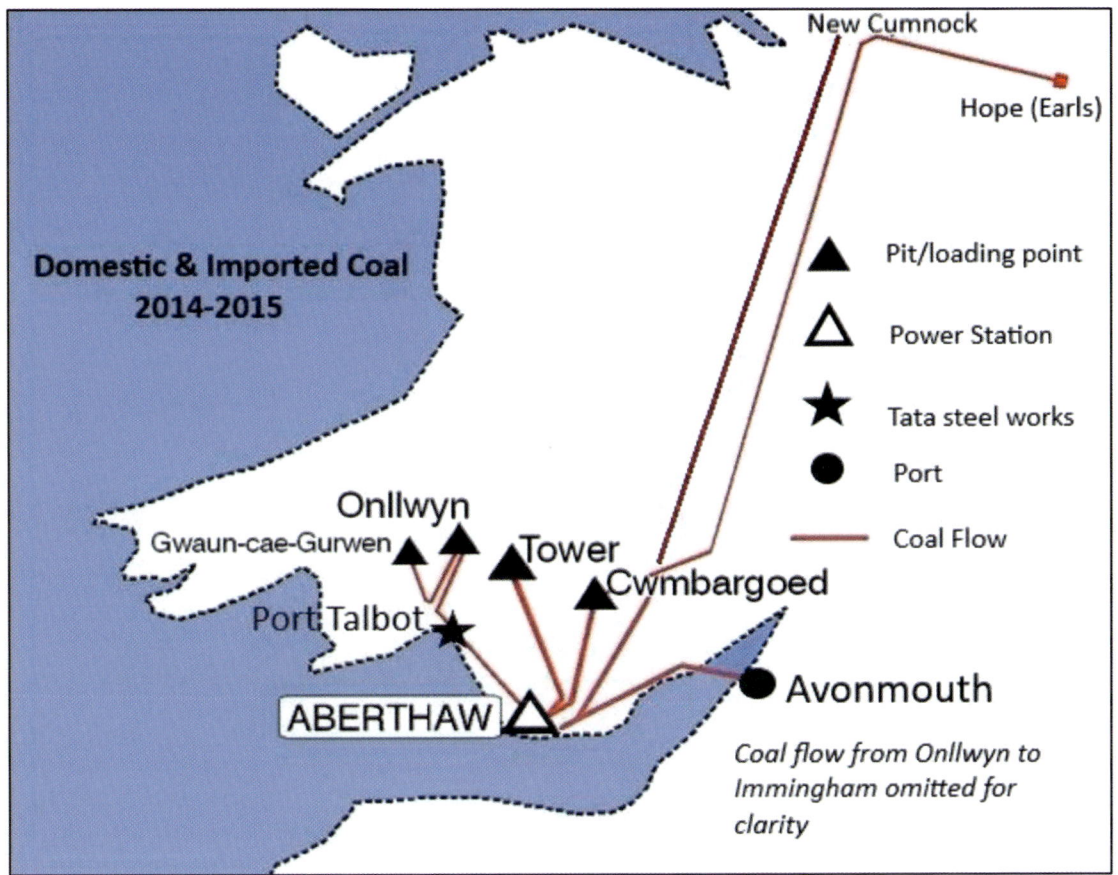

South Wales Coal Flows, 2014–15 C/O Freightmaster Publishing.

Coal despatched from Gwaun-Cae-Gurwen to Onllywn was an important flow until it ceased in May 2019. There is still a once-weekly train from Onllwyn to Immingham, which will discontinue in the next couple of years as the reserves at Nant Helen come to an end.

There is still demand for Welsh coal at Port Talbot steel works and Hope cement works, but continuation of the coaling operation at Cwmbargoed after 2022 is subject to planning permission by the local council.

The table below shows the coal flows for the period 2014–15 when rail traffic was relatively busy, largely due to demand at Aberthaw. By comparison, there was a dramatic fall from this position from 2017 onwards when Aberthaw stopped burning Welsh coal and only generated power on an intermittent basis until its closure in 2019. That said, although there continues to be demand for Welsh coal in the steel and cement manufacturing industry, the future outlook for these flows is uncertain.

Origin	Destination	FOC
South Wales Coal Flows 2014–15		
New Cumnock	Aberthaw	DBC
Cwmbargoed	Aberthaw, Port Talbot Grange, Hope Earles Sidings	DBC & FL
Onllwyn	Aberthaw, Immingham, Mossend	DBC & FL
Tower	Aberthaw	DBC & FL
Gwaun-Cae-Gurwen	Onllwyn	DBC
Avonmouth	Aberthaw	DBC, FL & Colas
Newport Docks	Aberthaw	DBC
Redcar	Port Talbot (coke)	DBC & FL
North Blyth	Port Talbot	DBC
South Wales Coal Flows 2018–20		
Cwmbargoed	Port Talbot Grange, Hope Earls Sidings, Scunthorpe	DBC & FL
Onllwyn	Immingham, Mossend, Scunthorpe*	DBC
Gwaun-Cae-Gurwen	Onllwyn, Immingham**	DBC
York/North Blyth	Port Talbot Grange	FL

FOC – Freight Operating Company; DBC – DB Cargo; FL – Freightliner; ** short-term only;
* new flow

Chapter 3
Cwmbargoed

Cwmbargoed is situated at the head of the Taff Bargoed Valley, approximately 4km east of Merthyr Tydfil and 4.5km north of the village of Bedlinog. The Cwmbargoed coal disposal point (DP) is owned and run by Merthyr (South Wales) Ltd (MSW) who operate the nearby Ffos-y-Fran opencast mine. Coal is brought in from the open pit to the wash plant (washery) for processing and subsequent despatch by train from the railhead located at the DP.

The DP was first commissioned by the National Coal Board in 1957 to blend and prepare local coals for distribution to customers throughout the UK and overseas. Today, the plant produces around 100,000 tonnes of washed coal per annum and a further 150,000 tonnes is produced from the barrel wash treatment of inferior coal seams and old workings. The washery also has a heavy media separation facility to produce high-grade coal for metallurgical use.

The Ffos-y-Fran mine is the UK's largest opencast coal operation. Coal reserves of 10.8 million tonnes from around 30 coal seams from the lower part of the coal measures were originally identified for surface extraction of between 0.8 to 1 million tonnes per annum. Around 8 million tonnes of coal has been mined to date. Planning permission to mine coal at Ffos-y-Fran expires in 2022. Additional reserves exist at the nearby Nant Llesg deposit.

There has been a long history of coal and iron ore mining in the area, the legacy of which has resulted in significant environmental damage. Under the initiative of Mid Glamorgan County Council, in partnership with the Welsh Development Agency, the East Merthyr Land Reclamation Scheme was launched in the late 1980s by the Secretary of State for Wales. The scheme stipulated that as a condition of the recommencement of coal mining in the area, mining companies would have to make good and restore land that had previously been mined and disturbed. The revenues from surface mining operations would be used to pay for the reclamation.

MSW is responsible for the final phase of the reclamation schemes, the 'Ffos-y-Fran Land Reclamation Scheme', which is the largest of the programmes, embracing an area of 467 hectares. The work requires the removal of old shafts, adits and mine workings, as well as the removal of fly tipping, including many burnt-out and stolen cars, from the site. It is estimated the reclamation will be completed by the end of 2024.

There was significant opposition to the mine. Following a public inquiry, however, the Welsh Assembly finally gave approval for the mine and associated reclamation scheme to go ahead in April 2005, subject to 59 planning conditions. After a hiatus of coal mining at Cwmbargoed for many years, the operation commenced in 2008 with the creation of 200 jobs.

Condition 37 of the planning conditions stipulated that 'no coal should be transported from the site except to the Cwmbargoed Disposal Point for onward transmission by rail in the interests of highway safety, residential amenity and sustainability'. Though a small proportion of coal is transported by road for the domestic market (mainly for use in steam trains), most of it leaves via the railhead from the DP. MSW estimate rail transportation will save up to 42 million lorry miles over the life of the mining operation and have an agreement with Network Rail for a maximum of 30 trains per week from the DP.

The coal is taken from the DP along the Cwmbargoed branch line to Ystrad Mynach, where it joins the Rhymney line to Cardiff and connects to the Vale of Glamorgan Line (VOG) to reach Aberthaw

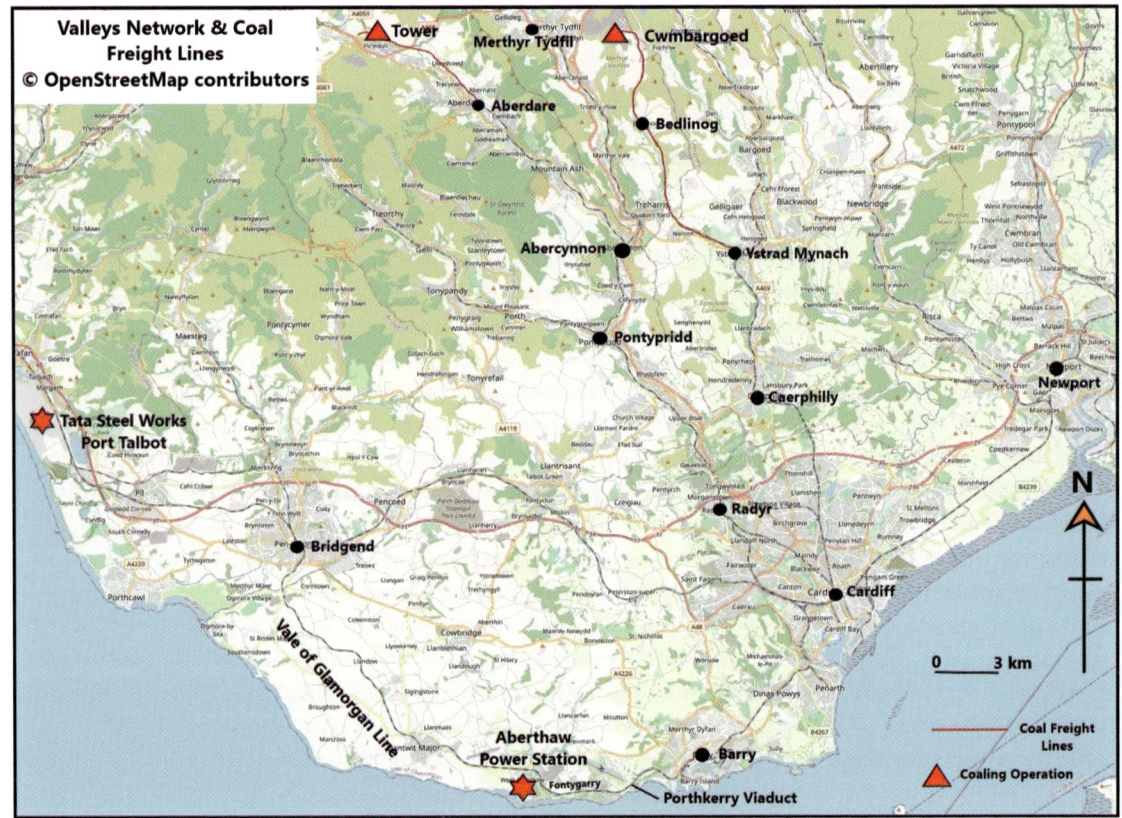

power station. Although Aberthaw 'B' closed to Welsh coal traffic in 2017, the VOG line is still used for the coal flows to Port Talbot Grange Sidings via Bridgend for Tata steel works, and to Hope (Earles Sidings) in Derbyshire, via Margam. On rare occasions, trains are diverted along the South Wales main line.

The former Rhymney and GW joint line was built in two stages; the section between Ystrad Mynach and Nelson & Llancaiach in 1871 and the section to Dowlais in 1876. The line served passenger traffic to Nelson & Llancaiach, Trelewis, Bedlinog, Cwmbargoed and Dowlais Cae Harris stations, but closed in 1964 when it became victim to the Beeching Act. It also used to serve collieries along the Taff Bargoed Valley and was used to bring imported iron ore from Cardiff (to replace locally mined iron ore) for the iron foundries at Dowlais. After withdrawal of passenger services in 1964, the line was singled throughout.

The steeply graded line to the DP runs for 14.5km (9 miles) from Ystrad Mynach. It has continuous gradients of between 1 in 145 (6.9mm/m) at Ystrad Mynach station, 1 in 220 (4.5mm/m) at Nelson, and for approximately 11.3 kilometres (7 miles) thereafter gradients of between 1 in 40 (25mm/m) and 1 in 44 (23.8mm/m) to Cwmbargoed. It is reported to climb some 200 metres in the last 6km section to the summit of the line at the DP.

The current speed on the line is 24kph (15mph) for the first 4km and 32kph (20mph) for the remaining 11km. The branch was previously under the control of Ystrad Mynach signal box, which had a splendid array of semaphore signals until they were replaced in September 2013 by LED lights and a remote control and tokenless block system, operated from the Cardiff Rail

Operating Centre. The branch is signalled as one section from Ystrad Mynach to Cwmbargoed. There is a loop, just west of the DP, used for locomotives to run round, which can accommodate two trains at the same time. A single track branches off to the DP where the trains push back to load up with coal.

For many years, Aberthaw power station was the primary customer for the finished coal product from Cwmbargoed, with up to 83 per cent of the coal sent to the power station. Ffos-y-Fran was considered a strategic coal reserve of prime importance to the UK energy balance due to the high-quality, low-volatile, bituminous coal which was ideal for the plant. The coal was considered more environmentally friendly than that mined elsewhere in the UK and could also be used to replace imported coal without putting at risk the existing market for coal from Tower and other South Wales mining operations.

Wednesday 9 January 2008 witnessed the first coal train to be loaded at the DP for many years. No. 66078 entered the reception sidings and pushed its wagons back to the washery for loading. The train then departed for Aberthaw power station with its load of around 1,500 tonnes of coal. This was a trend that continued until 2016, with at least two trains a day (Mondays to Saturdays). EWS, later DB Cargo, was the freight operating company (FOC) for many years until its last train departed on 31 March 2016, with No. 66152 in charge.

At the beginning of April 2016, Freightliner replaced DB Cargo as the FOC, and signed a five-year contract to move coal to Aberthaw, but it was short lived, lasting just one year due to the decision by the operating company, Rheinisch-Westfälisches Elektrizitätswerk (RWE), to stop using Welsh coal at Aberthaw. This was in response to the European Court of Justice who ruled that emissions of toxic nitrogen oxides (NO_x) from burning Welsh coal at Aberthaw were too high. Consequently, the stricter environmental regulations on emissions, coupled with the decision by RWE in 2017 to generate power only at peak times, meant the cessation of the supply of Welsh coal to Aberthaw. The final Freightliner train left on 15 March 2017 with 66/5 No. 66511 in charge.

Whilst this was the end of the regular coal flow to Aberthaw, six more trains operated by Freightliner in late September/early October 2018, made the journey to the power station with coal for blending purposes.

Though the loss of the Aberthaw contract must initially have come as a blow to MSW, the operation has continued relatively unscathed due to Tata Steel's demand for coal. Port Talbot steel works has been an important customer for the coal for many years and replaced Aberthaw as its principal customer in 2017. Metallurgical coal (coking coal) is a vital ingredient in the iron ore smelting part of the steel-making process, so MSW invested £10 million in a new state-of-the-art processing facility at the washery in 2015 to produce coal suitable for use in the high-pressure injection furnaces at Port Talbot.

From the start of operations, DB Cargo was the sole FOC for Tata Steel. However, Freightliner also ran trains to Port Talbot Grange in November and December 2016. This trend continued until May 2019 with both freight operating companies operating usually one, but sometimes two, trains a day in response to the increase in Tata's consumption requirements for coal. The contract has since reverted back to DB Cargo who is now the sole FOC at Cwmbargoed. Depending on requirements, DB Cargo sometimes runs a once-weekly train (Wednesday only) to British Steel's Scunthorpe works.

Another important coal flow is to Earles Sidings in Derbyshire for use at Breedon cement works. DB Cargo has been the sole FOC for this flow and trains usually run once, and sometimes twice, weekly (Monday/Friday only). Unlike the other coal flows, these trains comprise a shorter set of 18 HTA wagons (or fewer), a function of the short branch line to the cement works where the wagons are split six at a time. Another unusual feature is the use of BYA covered wagons (steel-coil wagons) utilised as screw coupling adaptors.

| Cwmbargoed Coal Flows, 2019–20 ||||||
Mondays–Saturdays	Train	Time	Headcode	FOC
	Margam–Cwmbargoed	05:25	4C93	DBC
	Margam–Cwmbargoed	08:36	4C94	DBC
	Cwmbargoed–Port Talbot Grange/Margam	14:00	6C93	DBC
	Cwmbargoed–Port Talbot Grange/Margam	17:47	6C94	DBC
W/O	Cwmbargoed–Scunthorpe	17:47	6E09	DBC
M/FO	Hope Earles Sidings–Cwmbargoed	23:58	4V01	DBC
M/FO	Cwmbargoed–Hope Earles Sidings	07:47	6M77	DBC

FOC – Freight Operating Company; DBC – DB Cargo; M/FO = Mondays/Fridays only; W/O = Wednesdays only

Since the start of operations in 2008, Class 66 locomotives have been used exclusively to haul the coal out of Cwmbargoed. There have, however, been one or two exceptions to this rule in recent years when Class 60 locomotives have made a rare appearance. On the 27 June 2014, No. 60065, *Spirit of Jaguar*, was the first Class 60 to arrive at Cwmbargoed since 1998. It was followed shortly after by No. 60001, which arrived on 8 September 2014. Both locomotives originated from Earles Sidings to supply coal to Breedon (Hope) cement works.

Despite the presence of additional coal reserves at the nearby Nant Llesg deposit, near Fochriw west of Rhymney, with potential for extraction of a further 6 million tonnes of coal, it remains uncertain if the coal will ever be mined as planning permission was rejected by Caerphilly council in 2015. The decision was based on the grounds of visual impact. A subsequent planning appeal by MSW was rejected.

Whether the line will see coal freight beyond 2022 is contingent on planning permission for mining to continue. Studies have been commissioned by Merthyr Tydfil County Borough Council (The Sewta Rail Strategy 2013) to look into the feasibility of extending the passenger rail service from Ystrad Mynach along the branch to Bedlinog via Nelson and Trelewis, including a service further north to Dowlais Top, north-east of Merthyr Tydfil. Therefore, the line may survive, with or without mining.

On 29 October 2013, No. 66114, working the 6C93 service with 21 loaded HTA coal hoppers destined for Aberthaw power station, slowly moves down the Taff Bargoed Valley at the start of its journey.

Just catching the sunshine, No. 66083, with 21 HTA hoppers laden with coal destined for Aberthaw power station, waits for its path to depart Cwmbargoed on 14 November 2014. The Ffos-y-Fran opencast pit is situated behind the train.

No. 66076 slowly makes it way down the Taff Bargoed Valley on the 6C93 to Aberthaw on 31 October 2014. The train is passing the former Nant-y-Ffin No2 small coal mine, which was abandoned in the mid 1980s. Some derelict buildings, including the old winding house, can still be seen at the site and remnants of the former mine infrastructure, including the coal discharge chute, can be seen to the right. The coal tip in the distance is from mine operations at Ffos-y-Fran.

A typical scene at Cwmbargoed with light and shadow playing on the bleak landscape sees No. 66083 depart the DP with another load of coal destined for Aberthaw power station on 14 November 2014.

Two's Company! On 16 January 2015, with snow-clad hills in the background, No. 66183 makes its way along the sidings at Cwmbargoed prior to pushing back its HTAs to the DP (out of sight) to load coal for its onwards journey to Hope in Derbyshire (Earles Sidings) via Margam on the 6M77. Meanwhile, No. 66015 on the left, having loaded with coal for Aberthaw power station on the 6C95, waits for its path to depart.

No. 66083 heads down the Taff Bargoed valley with 21 HTA hoppers laden with coal destined for Aberthaw power station (6C93) on 14 November 2014. The large coal tip in the background is from the Ffos-y-Fran pit, the UK's largest opencast coal mine.

On 14 November 2014, No. 66080 negotiates the steep climb to Cwmbargoed along the Taff Bargoed Valley, passing the site of the former Taff Merthyr colliery, seen to the left of the train, which closed in 1992. The train has originated from Earles Sidings in Deryshire (4V01) and will load with coal at the DP for the return journey to Hope (Earles Sidings) via Margam.

On a cold January winter's day (14 January 2015), No. 66183 departs Cwmbargoed for Aberthaw power station (6C95).

On 16 January 2015, No. 66232, seen at the head of the Taff Bargoed Valley in South Wales, climbs the steep gradient to Cwmbargoed on the 4V01 service to load with coal for the return journey to Hope (Earles Sidings) in Derbyshire via Margam. The line climbs some 200 metres in the last 6km section to Cwmbargoed, where this picture was taken.

DB Cargo 'Shed' No. 66183, passes a large spoil tip associated with the Ffos-y-Fran open-pit mine, as it descends the Taff Bargoed hauling around 1,500 tonnes of coal on the start of its journey to Aberthaw power station (6C95) on 14 January 2015.

No. 66063, seen at the head of the Taff Bargoed Valley in South Wales, negotiates the steep climb to Cwmbargoed on the 4C95 coal empties from Margam on 15 July 2015. The steeply graded 11km branch from the Rhymney line at Ystrad Mynach has continuous gradients of between 1:40 and 1:49 and climbs 200 metres in the last 6km section to Cwmbargoed.

On 2 November 2015, amid the warm autumn colours and splendid scenery, No. 66011 makes its way down the Taff Bargoed Valley with the 14:00 (6C93) Cwmbargoed–Aberthaw loaded coal.

No. 66136 is seen at the end of its journey just past the Cwmbargoed DP, working the 4V01 on 15 February 2016. The train will push back its shortened consist of 18 HTA wagons to load with coal for the return journey to Hope (Earles Sidings) in Derbyshire (6M77).

The steep grade of the line to Cwmbargoed, shown on 15 February 2016. No. 66136, seen at the head of the Taff Bargoed Valley, negotiates the steep climb to Cwmbargoed on the 4V01 coal empties from Hope (Earles Sidings) in Derbyshire.

DB Cargo 'Red Shed' No. 66085 passes through Bedlinog along the steeply graded line to Cwmbargoed with coal empties from Hope (Earles Sidings) on the 4V01 on 18 May 2018.

No. 66080 is seen loading up with coal at the Cwmbargoed DP for the onwards journey to Hope (Earles Sidings) in Derbyshire (6M77) on 14 November 2014. The train is unusual with its short consist of just 12 HTA wagons, a function of the short branch line to the cement works at Hope in Deryshire.

A view of Ystrad Mynach South Junction where the line to the left branches off to Cwmbargoed from the Rhymney line double track. It was taken from Ystrad Mynach signal box on 7 August 2013, when the line was still semaphored and controlled by the signal box.

On a cold, dark, damp, misty winter morning (18 December 2019) at Ystrad Mynach South Junction, No. 66020 enters the branch line to Cwmbargoed with 21 empty HTA wagons from Margam (6C93) for loading with coal for the return trip to Port Talbot. The branch used to be controlled by Ystrad Mynach signal box using a token system but this changed in September 2013, when the old signalling infrastructure (semaphores) was replaced by LED lights and a remote control and tokenless block system, operated from the Cardiff Rail Operating Centre.

On 1 September 2016, Freightliner 'Shed' 66/5 No. 66546 departs Cwmbargoed on the 6C93 with 20 HXA wagons in tow, laden with coal for Aberthaw power station. Meanwhile, the HTA wagons behind No. 66140 are being loaded with coal at the wash plant for the later 6C94 departure to Port Talbot Grange Sidings.

No. 66126 heading 21 empty HTA wagons arrives at Cwmbargoed from Port Talbot (4C93) on 4 July 2018 to collect coal for the return journey to the steel works.

No. 66126 heading 21 empty HTA wagons arrives at Cwmbargoed from Margam on 4 July 2018; it is seen past the DP on the 4C93. The train will edge forward past the Freightliner train and then push back its wagons to the DP to be loaded up with coal. Meanwhile, Freightliner 'Shed' 66/5 No. 66529 awaits to depart Cwmbargoed for Port Talbot Grange Sidings with 20 HHA wagons loaded with bituminous coal for use at Tata steel works (6C93).

Head to Head! Freightliner 'Shed' 66/5 No. 66546, in charge of 21 HTA wagons laden with coal, awaits its path for Aberthaw power station (6C93) on 1 September 2016. Meanwhile, DB Cargo 'Shed' No. 66140 loads with coal at the wash plant for the later departure to Port Talbot Grange Sidings on the 6C94.

No. 66197 heads down the Taff Bargoed Valley working the 6C93 Cwmbargoed–Port Talbot Grange Sidings on 10 October 2016. Meanwhile, an unidentified Class 66 heads the 6M77 Cwmbargoed–Hope (Earles Siding) service and loads up with coal at the wash plant.

On 10 October 2016, No. 66197 slowly makes its way along the eastern flank of the Taff Bargoed Valley heading south towards Ystrad Mynach, where it will join the Rhymney line for the onward journey to Cardiff and Port Talbot on the 14:00 (6C93) Cwmbargoed–Port Talbot Grange Sidings. Three collieries were in production in recent times not far from this location – Deep Navigation, Taff Merthyr and Trelewis Drift. All three mines closed in the early 1990s, following which there was considerable rehabilitation of the landscape, which now shows little evidence of the old coal mines.

The russet-brown bracken dominates the Taff Bargoed Valley in this scene as, on 3 December 2016, Freightliner 'Shed' 66/9 No. 66952 slowly descends down the valley with its cargo of 1,500 tonnes of coal destined for Port Talbot Grange Sidings (6C93). At this time, Freightliner were assisting DB Cargo with the coal flow to Port Talbot Grange.

No. 66031, with 18 empty HTA wagons in tow, negotiates the steep climb towards Cwmbargoed through the village of Bedlinog, working the 4V01 from Hope (Earles Sidings, Derbyshire) on 5 January 2017. It's a steep walk to get to this point from Bedlinog but well worth it with the stunning views of the valley.

Cwmbargoed

Two forms of energy, side by side. 66/5 No. 66570 departs Cwmbargoed on the daily 6C94 working to Port Talbot Grange Sidings. It is seen on 27 February 2019, passing the solar panel farm, which was installed in 2017.

On 5 March 2019, 66/5 No. 66570 passes through Trelewis as it climbs the Taff Bargoed Valley with coal empties from East Usk Junction (4C94), destined for Cwmbargoed. Behind the train, on the right, is the site of the former Deep Navigation Colliery, which closed in 1991. It was one of the few profitable NCB mines in South Wales but geological issues ultimately led to its closure. The train will return to Port Talbot on the 6C94 with around 1,500 tonnes of coal.

On 17 July 2019, No. 66055 pushes back a rake of 21 HTA wagons towards the DP at Cwmbargoed (4C94) for loading with coal for the onward journey to Port Talbot Grange. Meanwhile, No. 66103 is fully loaded ready to depart for Port Talbot Grange Sidings on the 6C93.

Freightliner 66/5 No. 66518 passes old mine workings of the Nant-y-Ffin No2 drift mine (abandoned in 1986) on 1 September 2017 as it descends the Taff Bargoed Valley to Port Talbot Grange Sidings with loaded coal from Cwmbargoed (6C93). Slag heaps from coal mining at Ffos-y-Fran open pit, which feeds the disposal point at Cwmbargoed, can be seen in the background, left of the train.

The coal trains to Cwmbargoed usually haul 20 HTA wagons (DB Cargo) or 20 HHA wagons (Freightliner). The exception to the rule are those trains originating from Hope (Earles Sidings) in Derbyshire which have a shorter set of 18 HTA wagons (or fewer), a function of the short branch line to the cement works where the wagons are split six at a time. In this 14 November 2014 scene, No. 66080 has a short consist of just 12 HTA wagons, which are being filled with coal at the DP for the return journey to Earles Sidings (6M77).

Surrounded by golden autumnal colours, on 1 November 2017, 66/5 No. 66525 heads the 14:00 (6C93) Cwmbargoed–Port Talbot Grange loaded coal train as it descends the Taff Bargoed Valley towards Ystrad Mynach, where it will join the Rhymney line for the onwards journey.

On 12 December 2017, No. 66158 arrives at Cwmbargoed on the 4C94 from Margam with 21 empty HTA coal hopper wagons to load with coal for the return journey to Port Talbot. The locomotive will push back the wagons to the Disposal Point on the right to load up.

Winter in the Valleys as, on 12 December 2017, Freightliner 'Shed' No. 66/5 66511 slowly heads the 6C93 down the snow-dusted Taff Bargoed Valley with around 1,500 tonnes of coal from the Ffos-y-Fran open pit, the UK's largest opencast coal mine. The train is on its way to Port Talbot Grange Sidings where the coal will be used for Tata steel works.

On 21 August 2019, DB Cargo 'Shed' No. 66014 loads up with coal to take to Port Talbot Grange on the 6C94. The conveyor belt above the train is discharging washed and blended coal to the stockpile. The stockpiled coal in the foreground is from the Ffos-y-Fran open pit for processing at the washery.

On 5 March 2019, 66/5 No. 66570 climbs up the Taff Bargoed Valley with empty hoppers in tow, en-route to Cwmbargoed from Margam on the 4C94.

One of the last Freightliner trains to Cwmbargoed sees 66/5 No. 66517 charging through Caerphilly Station with coal empties from Margam (4C93) on 23 April 2019.

Freightliner 'Shed' 66/5 No. 66529 awaits to depart on the 14:00 (6C93) Cwmbargoed–Port Talbot Grange loaded coal on 4 July 2018.

On 21 August 2019, DB Cargo 'Shed' No. 66014 loads up with coal at the DP for the return journey to Port Talbot steel works on the 6C94.

Slowly does it! No. 66070 slowly makes its way down the Taff Bargoed Valley through Bedlinog on a wet autumnal afternoon (17 October 2019) with 21 loaded HTA wagons, carrying around 1,500 tonnes of coal, in tow. This is the 14:00 (6C93) Cwmbargoed–Port Talbot Grange daily working.

Viewed on 7 March 2019 from old coal tips near the small town of Treharris, 66/5 No. 66570 heads down the Taff Bargoed Valley, working the 14:00 (6C94) Cwmbargoed–Port Talbot Grange loaded coal. The valley floor used to be the site of the Taff Merthyr Colliery, which shut down in 1992 and was the penultimate colliery to close in South Wales. It is now occupied by an indoor rock-climbing facility.

On a dull, dreary, dank day (7 March 2019), No. 66040 arrives at Cwmbargoed from Margam on the 4C94 with empty hoppers. The train will push back to the disposal point behind it to the left (out of view) to load up with coal for the return trip to Port Talbot (6C94).

No. 66158 heads through the village of Bedlinog as it climbs up the snow-dusted Taff Bargoed Valley towards Cwmbargoed, working the 4C94 coal empties from Margam on 12 December 2017.

Having joined the single-track line branch at Ystrad Mynach on 13 February 2019, No. 66186 makes its way up the Taff Bargoed Valley with coal empties to Cwmbargoed from Margam (4C94). The buildings below in the valley, which are used for an indoor rock-climbing facility, occupy the site of the former Taff Merthyr Colliery, which closed in 1992 despite potential for another ten years of reserves. The train will load up with around 1,500 tonnes of coal and proceed to Scunthorpe (6E09).

On 18 February 2019, Freightliner 'Shed' 66/9 No. 66956 travels along the Taff Bargoed Valley, past the access point to the former Taff Merthyr Colliery, in charge of a loaded coal train from Cwmbargoed which is destined for Port Talbot Grange Sidings (6C94).

Cwmbargoed

Freightliner 'Shed' 66/5 No. 66508 joins the Rhymney line from the Cwmbargoed branch at Ystrad Mynach South Junction, on the 14:00 (6C93) Cwmbargoed–Port Talbot Grange Sidings on 25 March 2019.

Freightliner 'Shed' 66/5 No. 66561 approaches Trelewis, working the 14:00 (6C93) Cwmbargoed–Port Talbot Grange loaded coal on 28 March 2019.

On a lovely summer afternoon (17 July 2019), DB Cargo 'Shed' No. 66055 loads up with coal at the Cwmbargoed DP for the 17:47 (6E09) to Scunthorpe. Meanwhile, DB Cargo 'Shed' No. 66103 passes a solar energy farm having just left the DP on the 14:00 (6C93) working to Port Talbot Grange Sidings, hauling 21 loaded HTA wagons.

In typically inclement weather, on 15 August 2019, DB Cargo 'Shed' No. 66065 pushes back a rake of empty HTA wagons, on the 4C93 from Margam to the disposal point at Cwmbargoed, for loading with coal.

On 13 August 2019, dump trucks from the nearby Ffos-y-Fran open pit haul coal to the washery for washing and screening. Meanwhile, having just loaded with coal, DB Cargo 'Shed' No. 66104 heads out of the washery to the loop sidings at Cwmbargoed. The locomotive will run around its trains ready for the 14:00 departure to Port Talbot Grange Sidings on the 6C94.

Freightliner 'Shed' 66/5 No. 66529 waits to depart Cwmbargoed with loaded coal on the 14:00 (6C93) to Port Talbot Grange Sidings. Taken with permission from the mine management on 4 July 2018, this image shows Ffos-y-Fran opencast pit in the background, left of the coal tips.

On 29 January 2020, 66/5 No. 66151 slowly descends the Taff Bargoed Valley as it passes through the village of Bedlinog, with 1,500 tonnes of coal in tow destined for Margam (Port Talbot) on the 6C94 from Cwmbargoed.

Freightliner 'Shed' 66/5 No. 66531 makes its way along the Taff Bargoed Valley, heading south near the former Taff Mertrhyr Colliery, working the 6C94 Cwmbargoed to Port Talbot Grange Sidings loaded coal train on 13 February 2019.

On 29 January 2020, DB Cargo 'Shed' No 66169 makes the steep climb through the site of the former Bedlinog station in the Taff Bargoed Valley, hauling 21 empty HTA wagons to Cwmbargoed on the 4C94 from Margam. The train will return on the 6C94 with around 1,500 tonnes of coal for use at Tata steel works.

In winter sunshine, No. 66169 passes through the village of Bedlinog in the Taff Bargoed Valley, hauling 21 HTA wagons up the steep gradient to Cwmbargoed on the 4C94 from Margam on 29 January 2020.

On 29 January 2020, DB Cargo 'Shed' No 66151 waits at Cwmbargoed for its path on the 6C94 departure to Margam with around 1,500 tonnes of bituminous coal that will be used at Port Talbot steel works.

On 5 February 2020, No. 66051 makes its way down the single-track line to Ystrad Mynach, at the lower end of the Taff Bargoed Valley, towards Ystrad Mynach Junction, working the 6C93 from Cwmbargoed with a loaded coal train for Port Talbot Grange.

Journey's End! On 16 October 2019, having arrived at Margam Knuckle Yard from Cwmbargoed on the 6C93, with around 1,500 tonnes of coal in tow, DB Cargo liveried 'Shed' No. 66019 takes the train on the next short hop to Grange Sidings at Port Talbot steel works to unload its coal.

DB Cargo 'Shed' No. 66077, working the 6C93, brings another load of coal from Cwmbargoed to Port Talbot steel works on 19 September 2019.

On 12 December 2017, Freightliner 'Shed' 66/5 No. 66511 slowly heads south down the picturesque, snow-dusted Taff Bargoed Valley on the 6C93 from Cwmbargoed, hauling around 1,500 tonnes of coal destined for Port Talbot steel works. The coal trains from Cwmbargoed and Onllwyn in South Wales are amongst the last to haul coal still mined in the UK.

An important coal flow from Cwmbargoed in South Wales is to Earles Sidings in Derbyshire for use at Breedon cement works. Operated by DB Cargo, trains usually run once, and sometimes twice, weekly (Monday/Friday only). In this 17 June 2017 scene, a Class 20 locomotive (formerly No. 20906, now No. 3), owned by Breedon Cement Ltd, has just come off the short branch to the cement works on the left, and is pushing the last 6 HTA wagons at Earles Sidings to make a rake of 18 wagons for the next trip to Cwmbargoed.

Chapter 4
Tower

Situated near the villages of Hirwaun and Rhigos, north of Aberdare in the Cynon Valley, Tower Colliery has a unique and fascinating history. It became famous for being the last deep underground colliery in South Wales to close. After the end of the 1984/85 miners' strike, which lasted 12 months and affected the whole of the UK, the Conservative government authorised British Coal to close most of the UK's deep pits as they were deemed to be uneconomic. Tower, however, managed to survive for a few more years, mainly as a result of investment in new underground mine development, but its days were numbered. It combined with Maerdy colliery, effectively becoming one mining complex raising all Maerdy's coal production until that pit's closure in December 1990. However, its reprieve was short lived. On 22 April 1994, British Coal closed Tower Colliery stating the pit was no longer an economic operating mine.

One major geological advantage of Tower was that it produced high-quality anthracite, as opposed to bituminous coal, which potentially paved the way for ensuring continuing markets. In addition, the pit was considered to have at least five years of reserves left. Therefore, rather than accept closure of the colliery, spurred by NUM branch secretary, Tyrone O'Sullivan, together with 239 miners, an offer was made to British Coal to purchase the colliery, so that production could continue. A price of £2m was agreed for the sale of the mine. Each miner contributed £8,000 from their redundancy package and collectively became joint owners (shareholders) of Tower under a new Company, Goitre Tower Anthracite Ltd.

On 2 January 1995, behind a brass band, the miners reopened the pit and ran it themselves as a going concern. It overwhelmingly defied the odds and ran at a profit for the next 13 years until its closure in 2008. It was the oldest working deep coal mine in Britain and the last of its kind to operate in the Valleys. Its final demise was due to the exhaustion of economic reserves at depth, coupled with geological issues.

That was not the end of the story though; an additional 6 million tonnes of unmined coal reserves in the upper levels of the coal bearing strata was identified. Furthermore, the coal was amenable to exploitation by open pit, a cheaper method of mining the coal. This meant the second major reprieve for the coal operation at Hirwaun.

In late 2011, Rhonnda Cynon Taff County Borough granted permission to exploit the coal by open pit in a joint venture between Hargreaves Mining and Tower Colliery Ltd – Tower Regeneration Ltd (TRL). The joint venture agreed a £42.5m finance package to fund the project. Employing 150 men, some of whom had worked in the deep pit, TRL commenced production in 2012. The new pit had an estimated 'mine life' of seven years with the production of around 0.8 to 1m tonnes per annum. The open pit was designed to reach a vertical depth of 165m, embracing an area of 200 hectares at the site of an old washery. With a coal price of £50 per tonne at that time, an annual turnover of around £50m was anticipated from the new venture, so it turned out to be more lucrative than the deep pit, which had a turnover of £18–26m. The joint venture was also responsible for the remediation and redevelopment of the site when mining ceased.

Despite the success of the TRL operation, production was cut short in 2016 as a result of the European Court of Justice ruling that emissions of toxic nitrogen oxides (NO_x) from Aberthaw were too high, even though the operating company at the power station, Rheinisch–Westfälisches

Elektrizitätswerk (RWE), maintained it was fully compliant with all permits to control emissions. The stricter environmental regulations on emissions, coupled with the decision that Aberthaw would only generate power at peak times in 2017 using higher volatile imported coal to reduce NO_x and sulphur emissions, meant the cessation of Welsh coal supply to the power station. Even though there was potential to extract a further 1.2 million tonnes of coal, this was the nail in the coffin for the TRL operation at Hirwaun.

The 7.5km-long single track from Aberdare to Tower was originally used to serve the collieries between Glynneath and Hirwaun and for passenger services to Hirwaun station, the latter ceasing in 1964 under the Beeching Act. Its importance as a vital link to take the coal out of Tower after the closure of passenger services ensured its survival for more than half a century later, until its closure in 2017. There is a short run-round loop at Tower but due to its short length, the trains would need to carry out an extended run-round manoeuvre to fill up the wagons.

The route from Aberdare is along the single track to Abercynon, and then through to Radyr via Pontypridd. From Radyr, the trains would follow the Cardiff City line to Penarth South Curve Junction for the onward journey along the Vale of Glamorgan line to Aberthaw. A few gaps in the half-hourly Arriva Trains Wales passenger services were required to allow the path for the trains, which could be looped at Stormstown, between Pontypridd and Abercynon, and Aberdare.

Aberthaw power station was the main customer for the coal mined from both Tower's deep pit and later from TRL's opencast operation. Trains ran twice (and sometimes three times) a day, with DB Cargo (formerly EWS), the principal freight operating company (FOC), running for some 20 years.

The first trains from the open pit started running in April 2012, hauling coal in 21 HTA wagons. This continued for the next four years until Freightliner replaced DB Cargo as the FOC on 1 April 2016, running trains with 20 HXA wagons. However, the contract was short lived, due to the early closure of TRL's operation in 2017. The final train to Aberthaw left on 24 February 2017, hauled by 66/5 No. 66519.

Tower Freight Coal Flows 2014 – Monday to Friday			
	Time	Head Code	FOC
Aberthaw–Tower	03:20	4C42	DBC
Tower–Aberthaw	10:54	6C45	DBC
Aberthaw–Tower	10:50	4C46	DBC
Tower–Aberthaw	18:17	6C47	DBC
Aberthaw–Tower	15:40	4C48	DBC
Tower–Aberthaw	22:35	6C49	DBC

FOC – Freight Operating Company; DBC – DB Cargo (Formerly DB Schenker)

For a short while in the summer of 2016, a handful of trial trains conveyed coal to Ketton Cement Works in the county of Rutland. The trains, hauled by DB Cargo, each comprised 36 loaded MEA wagons for the journey to Ketton Ward Sidings. This was the first and only time the line saw the use of MEA wagons at Tower. This was because Ketton could not accommodate HTA wagons due to the absence of unloading facilities at the cement plant.

The last trains to run from Tower went to Earles Sidings in Derbyshire for use at Breedon (Hope) cement works. DB Cargo was the FOC, running just one train a week on Thursdays, carrying the final tonnes of stockpiled coal from the TRL operation. It was a short-term working that temporarily replaced the coal flow from Cwmbargoed. The last train ran on 26 May 2017, marking the end of coal production at Tower and the last coal train in the Cynon Valley.

The single-track line from Tower is seen north of Aberdare Station on 4 August 2016. Virtually all the coal from Tower Colliery (mined by open pit in recent years) went to Aberthaw power station. However, in 2016, a handful of coal workings went to Ketton Ward Sidings with loaded MEA wagons, as seen in this image of No. 66076 at the front of the train (6E13).

On 3 February 2015, No. 66114 passes through Pontypridd, working the 6C45 Tower–Aberthaw, loaded with coal.

On 23 January 2015, No. 66199, working the 6C45 Tower–Aberthaw power station loaded coal train, is seen heading south at Morganstown near Radyr, north of Cardiff. On the right is Castle Coch, a 19th-century revival castle.

With the last of the evening's sun catching Castle Coch, 66/5 No. 66551 hauls 20 HXA wagons loaded with coal for Aberthaw power station on the 6C47 from Tower on 24 June 2016. Having just passed through Taffs Well, the train heads south along the Taff valley cut through Carboniferous limestone. Taffs Well has the distinction of being the only well in Wales associated with a thermal spring; the warm water is thought to be associated with a fault in the limestone.

No. 66025 passes through Taffs Well on the 6C45 from Tower with another load of coal for Aberthaw power station on 16 December 2014.

The last coal trains to leave Tower were hauled by DB Cargo to Earles Sidings in Derbyshire, on their way to Breedon (Hope) cement works, and ran just once a week. On a wet evening, No. 66116 powers the 6C47 through Pontypridd station en route to Newport ADJ and Earles Sidings on 31 March 2017.

On 12 May 2016, 66/5 No. 66548 heads south through Pontypridd station, working the 6C45 Tower–Aberthaw power station loaded coal service. The last train from Tower to Aberthaw ran on 24 February 2017, headed by FL 'Shed' No. 66519.

Freightliner 'Shed' No. 66/5 66509 rounds the curve as it approaches Pontypridd station with coal empties to Tower on the 4C46 from Aberthaw power station on 12 May 2016.

For a short while in the summer of 2016, a handful of trial trains from Tower conveyed coal to Ketton Cement Works in the county of Rutland. In this scene, No. 66076 climbs the incline on the approach to Abercynon with loaded MEA wagons, working the 6E13 on 4 August 2016.

An unidentified Class 66 locomotive works the 6C45 loaded coal to Aberthaw. It is seen passing through Aberdare in the Cynon Valley on a splendid autumn day (4 November 2013).

Abandoned and vandalised, the old Aberdare station is seen on 22 October 2015. The station opened in 1851 on the Great Western Railway Neath to Pontypool Line, and was last used on 13 June 1964, as the line was axed under the Beeching Act. The good news is that the building has recently been renovated as part of the new Coleg y Cymoedd Aberdare campus. In this scene, No. 66250 passes the catch points with empty HTA coal hoppers on its way to Tower from Aberthaw (4C46). The loaded train will return to the power station in the early evening.

On 11 December 2014, No. 66027 passes the abandoned signal box at Pontypridd Junction on the 6C45 Tower Colliery to Aberthaw power station with around 1,500 tonnes of coal in tow. The signal box survives for now due to its Grade II listed status, but no effort has been made to maintain the building and it is now derelict and boarded up.

No. 66085 approaches Abercynnon station on the 6C45 from Tower with coal destined for Aberthaw power station on 13 October 2015. The diamond crossover track configuration is particularly interesting. The line to the left serves Aberdare and continues to Tower. The line to the right serves Merthyr Tydfil. The track configuration was a result of a new island station at Abercynon, built in 2008. Originally, there were two separate stations at Abercynon (Abercynon North and Abercynon South), the south station was reconfigured to allow Aberdare services to use the station.

No. 66221 slowly makes its way along the Taff Valley on the approach to Pontypridd, taking loaded coal from Tower on the 6C45 to Aberthaw on 30 January 2015.

No. 66012 passes through Penrhiwceiber, a former coal-mining village and community in the Cynon Valley, working the 6C45 Tower–Aberthaw loaded coal service on 17 March 2014.

On 14 March 2014, No. 66019 waits at Aberdare for the path to proceed a coal freight to Aberthaw power station on the 6C45 from Tower. The train has travelled along the 7.5km single-track line from Hirwaun. On the right is the old Aberdare station.

No. 66027 passes the abandoned signal box at Pontypridd Junction on the 6C45 Tower to Aberthaw on 11 December 2014. Pontypridd signal box closed in 1998 but still survives at the junction due to its Grade II listed status. The design is by McKenzie and Holland and is a modified version of their Type 3 design, which had 230 levers. It controlled the Rhonnda line to the left and the Taff Vale line.

A horse bolts as No. 66120 powers through Morganstown on 20 April 2015, with another load of coal from Tower destined for Aberthaw power station (6C45). On the right is Castle Coch.

On 3 March 2014, No. 66026 heads north through Pontypridd station with coal empties for Tower on the 4C46. This junction was incredibly busy when coal mining flourished in South Wales, and freight trains provided coal from the Rhonnda and Aberdare (Cynon) Valleys to Newport, Cardiff and Barry Docks for export. The line to the left goes to Treherbert along the Rhondda Valley, which nowadays is only used for passenger services.

No. 66154 powers up as it passes the catch points at the old Aberdare station with the 6C45 loaded coal to Aberthaw from Tower on 22 October 2015.

No. 66076 heads along the banks of the River Taff as it approaches Radyr near Cardiff, working the 6C45 Tower to Aberthaw loaded coal on 4 November 2014.

No. 66206 rounds the curve at Pontypridd station as it heads south towards Cardiff with coal for Aberthaw (6C45) on 6 May 2014. Pontypridd station had the distinction of being the world's longest island platform due to the necessity to accommodate many converging railways lines on what became the 19-century hub of the South Wales Valley lines. In 2011, Network Rail refurbished the station, including the long canopy, so it now retains its original charm and character.

Chapter 5
Onllwyn

Located at the head of the Dulais Valley in the Neath Port Talbot county borough, Onllwyn lies near the former mining villages of Seven Sisters and Banwen. The area has a long history of coal mining dating back to 1841 when many mines were developed to exploit rich anthracite seams. In 1932, a wash plant (washery) to process coal from drift mines was established at Onllwyn and upgraded in the 1950s. Whilst the anthracite drift mines have long gone, coal processing at Onllwyn is still active to this day, run by Celtic Energy.

The Onllwyn distribution centre and washery is served by the freight-only line from Swansea Burrows that runs for approximately 26km (16 miles) via Neath. There is no direct route from the South Wales main line to Onllwyn, so all the trains run from the Swansea and District line to Swansea Burrows, a large railway yard that used to serve Swansea Docks, where the locomotives run round their train for the journey north to Onllwyn.

The line from Neath to Onllwyn was formerly part of the Neath & Brecon line. The concept of the Dulais Valley Mineral Railway to bring coal down from Onllwyn to the waterway at Neath received Royal Assent in 1862. However, John Dickson, who built the line, considered that for it to be successful it would need to be extended south to Swansea and north-east to Brecon. Permission was not granted for the extension to Swansea, but the line was able to continue to Brecon and arrangements were made with the Brecon and Merthyr Railway to use Brecon station. Accordingly, the company changed its name to the Neath & Brecon Railway, which was incorporated in 1863. The line opened as far as Onllwyn in 1863 and was completed to Brecon in 1867.

In later years, the Midland Railway acquired rights over the line and ran freight trains for many years, although it was expensive and difficult to operate, due to its remoteness. In 1922, the Neath & Brecon Railway was absorbed into the new Great Western Railway (GWR). After the acquisition by GWR, however, many freight trains were diverted to other routes and so the railway suffered, reverting to a quiet, remote rural line. Nevertheless, it was still an important artery for anthracite production in the Onllwyn area and continued to provide a steady flow of traffic to Swansea.

After the railways were nationalised in 1948, passenger numbers fell considerably, particularly north of Onllwyn, between Colbren Junction and Brecon. Hence, services were cut back, but miners continued to use the line between Neath and Colbren Junction. The sparse services, however, generally attracted few passengers and were withdrawn completely in 1962. Due to the continuing use of the washery at Onllwyn, the line survives to this day, though traffic has diminished substantially since 2019 due to the closure of East Pit. There is currently only one train operating per week to Immingham or Scunthorpe.

The line is largely single track, although there is a loop south of Neath & Brecon Junction to allow passing trains. At Neath & Brecon Junction the line branches to Cwmgwrach, which has not seen a coal train since 2013, when the Unity coal mine went into administration. A real gem at this location is the delightful, tucked away, little signal box situated at the end of the old Neath Riverside station, which closed in 1962. The signal box is a GWR box of GW5 design built in 1892. In 1957, it was recorded as having 14 lever frames. It's marvellous that the little box, which appears to have undergone quite a lot of renovation in recent times, still survives to this day, though with only one coal train a week currently using the line, its days are surely numbered. With the current paucity of freight, the signal box must be largely unmanned.

In 1998, Celtic Energy built a new, state-of-the-art, coal-processing plant (washery) at Onllwyn, where coal is washed, screened and sorted. The washery can process the production from all Celtic Energy's mining operations. It operates a heavy media drum separator for coal greater than 22mm and heavy media cyclones for smaller-sized coal fractions. These can run up to 24 hours per day, processing up to 26,000 tonnes per week of raw mined production and converting it into top-quality anthracite products. A froth flotation plant was constructed in 2005 as a facility to recover fine coal from washery slurry, which had previously been discarded as waste.

The plant produces five different sizes of products, embracing three different coal qualities: Black Diamond, Premium and Economy. It is also set up to produce a range of specialised sized coal products, to fulfil specific customer requirements.

During the past decade, Onllwyn has received coal from Nant Helen open pit by dump truck, Gwaun-Cae-Gurwen (GCG) in the Amman Valley by road and rail, and from Selar Pit near Glynneath by road wagon, which ceased in 2015. In 2019, the coal supply from GCG finished, but Onllwyn continues to be supplied from Nant Helen pit.

For many years, Aberthaw power station was Onllwyn's principal customer. DB Cargo ran one or two trains most days using Class 66 locomotives hauling coal in 21 HTA wagons. On Tuesdays and Thursdays, trains also ran to Immingham. In May 2016, Freightliner replaced DB Cargo

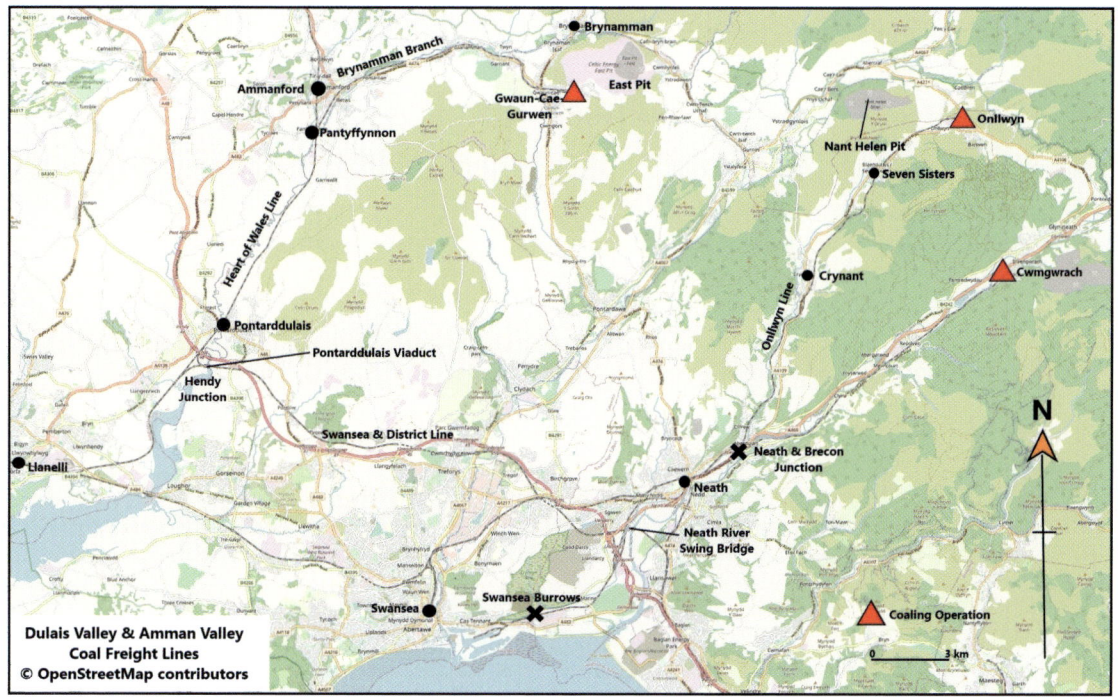

for the coal flow to Aberthaw, but the number of trains was reduced considerably. After many years of rail coal haulage, 9 September 2016 saw the last ever coal train to Aberthaw, headed by 66/5 No. 66552. Freightliner also occasionally worked a Monday-only train to Scunthorpe steel works.

Trains continued to run to Immingham after DB Cargo lost the Aberthaw contract. Until recently, there were two trains per week on Tuesdays and Thursdays, but this has now been reduced to just one on Tuesdays as a result of the winding down of operations at GCG. The trains either originate from Margam, bringing in empty HTA wagons for loading at Onllwyn via Swansea Burrows, or travel directly from Immingham in Lincolnshire. Coal Products Group, which has a briquette factory at the Port of Immingham, uses the coal to produce a whole range of high-quality smokeless fuels for the domestic and export markets.

Until recently, Onllwyn received rail-borne anthracite, mined at Celtic Energy's East Pit opencast coal mine in the Amman Valley, for processing from GCG. The first train ran in January 2012 when the line at Bryamman was reopened. DB Cargo was the FOC and they used Class 66 locomotives on the trains, which initially conveyed 25 MEA wagons twice a week to Swansea Burrows. This was later increased to 40 MEAs (and sometimes 45 MEAs) on a Tuesday/Thursday-only basis. The loaded MEA wagons would stay at Swansea Burrows overnight. A locomotive deployed from Margam would then take the train north to Onllwyn the next day. Two trains, each containing 20 MEA wagons, were necessary due to the trailing weights and gradients on the Onllwyn branch. The return journey was as a single train hauling all 40 empty MEA wagons back to Swansea Burrows. Class 66 locomotives from Margam were mostly utilised for this flow, although on rare occasions Class 60 locomotives were also operated.

In May 2019, the flow from GCG discontinued as the East Pit mine came to the end of its productive life. The final train, worked by No. 66171, departed for Swansea Burrows on 9 May 2019. The following day, the train was taken up in two loads to Onllwyn by No. 60091.

\multicolumn{5}{c}{Onllwyn Freight Flow 2018–19}				
Monday-Fridays	**Train**	**Time**	**Headcode**	**FOC**
T/ThO	Margam–Onllwyn	10:39	4O32	DBC
TO	Immingham–Onllwyn	10:39	4V32	DBC
T/ThO	Onllwyn–Immingham	08:36	6E09	DBC
W/FO	Swansea Burrows–Onllwyn	06:41	6O07	DBC
W/FO	Swansea Burrows–Onllwyn	11:45	6O08	DBC
W/FO	Onllwyn–Swansea Burrows	14:06	6O10	DBC

FOC – Freight Operating Company; DBC – DB Cargo; T – Tuesday; W – Wednesday; Th – Thursday; F – Friday; O – denotes 'only' e.g. ThO – Thursdays only

On rare occasions, coal is also despatched to Mossend, via Margam, using FCA/FYA twin- or FQA triple-container wagons. The trains, operated by DB Cargo, are the last rail-hauled containerised coal workings on the UK network. However, there have been no reported trains since the summer of 2019 so it is uncertain if the flow is still continuing. As of March 2020, a new coal flow to Scunthorpe operated by DB Cargo has commenced. This runs on Thursdays under the headcode 6E22.

It is worth mentioning that Onllwyn used to operate an industrial shunter, No. 08613, owned by RMS Locotec, to shunt the MEA wagons for unloading. Sadly, the shunter left in 2017, leaving the main line locomotives to take over running the wagons to the main pad.

The only pit currently providing coal to the washery is Nant Helen, which is situated approximately 2km west of the wash plant. It was cited as Celtic Energy's largest coal mining operation with initial reserves of around 3 million tonnes and an annual output of up to 400,000 tonnes. It is estimated that it has around 800,000 tonnes of reserves remaining, which will be used for the contracts to Immingham. As the pit is located near to the wash plant, coal is brought in by mine dump trucks, which also take the processing waste back to the pit for disposal.

Although the mine was mothballed in 2016 due to loss of demand, approval for the mine to extend operations from its original deadline of December 2018 to December 2021 was granted by Powys County Council in March 2019, with restoration work to be completed by June 2023. This is a favourable decision now that mining at East Pit has come to the end of its productive life. Whilst met with some opposition, it has meant a reprieve for operations at Onllwyn in the short-term, ensuring protection of jobs at the mine and washery for the next two to three years and, of course, survival of the freight line during this period.

The line to Onllwyn is unlikely to see a coal train again when mining finally ends at Nant Helen, but this does not necessarily mean the end of the line. A proposal has been made to build an international rail test centre when mining ceases at the open pit, for a cost of around £100m. The plan is to develop two electrified oval tracks for testing speeds of 110mph, with the option of a tunnel section and a full platform environment for testing and training. It, therefore, looks like the former Neath & Brecon line could survive for some time yet, despite the closure of Onllwyn in the next couple of years.

On 30 August 2018, No. 66095 loads up with coal at Onllwyn and will later depart to Immingham on the 6E09. The truck nearest to the camera is hauling coal from the nearby Nant Helen pit to the plant for washing, screening and grading. The truck travelling in the opposite direction is hauling waste from the washery for disposal at the pit.

No. 66070, destined for Immingham on the 6E09, gets ready to depart Onllwyn on 5 April 2016 with 21 loaded HTA hoppers. In the background is No. 08613, owned by RMS Locotec. It was used for shunting duties when coal was brought in from Gwaun-Cae-Gurwen, and for shunting container wagons but sadly is no longer used at Onllwyn.

No.66115 is seen on 26 January 2018 at the Onllwyn washery with the second train of coal for processing that day from Swansea Burrows on the 6O08. The coal (anthracite) is from Gwaun-Cae-Gurwen, near the East Pit opencast coal mine operated by Celtic Energy.

No. 66155 arrives at Onllwyn, on 26 January 2018, with a second load of coal from Swansea Burrows on the 6O08. The coal is from East Pit in the Amman Valley and was loaded at Gwaen-Cae-Gurwen.

No. 66155 travels light engine as it passes through Seven Sisters in the Dulais Valley, along the former Neath & Brecon line, on the way to Swansea Burrows to collect another load of coal for the washery at Onllwyn on 26 January 2018. Seven Sisters is important historically for coal mining and is cited as one of the richest sources of coal (anthracite) in the UK. Mining in the village ceased in 1960 when the men were transferred to the nearby Blaenant Colliery, which closed in 1990. Mining still continues at Nant Helen open-pit mine (behind the coal tip in the background) from which coal is trucked to the washery at Onllwyn.

A view of the washery at Onllwyn taken on a cold January morning (26 January 2018) at around 07:30. No. 66115 has brought in the first of two loads of coal, comprising 20 MEA wagons, from Swansea Burrows on the 6O07. The locomotive will return light engine to collect the second load of 20 MEA wagons and then return to Swansea Burrows with the complete rake of 40 empty MEA wagons.

On March 29 2019, DBC 'Shed' No. 66023, hauling loaded coal containers destined for Mossend via Margam on the 6O10 Onllwyn to Swansea Burrows, is seen passing through Seven Sisters in the Dulais Valley on the former Neath & Brecon line. This is amongst the last of the containerised coal workings on the network.

No. 66095 heads down the Dulais Valley through Seven Sisters, working the 6E09 Onllwyn–Immingham loaded coal train on 30 August 2018.

On 26 January 2018, No. 66155 gets ready to depart Onllwyn washery with 40 empty MEA wagons destined for Swansea Burrows (6O10), having discharged its coal for processing at the washery on two earlier workings (6O07 and 6O08).

No. 66221 loads up with processed coal at Onllwyn for the 6E09 journey to Immingham on 31 August 2017. The coal (anthracite) goes to a briquette factory at the Port of Immingham, where a range of high-quality smokeless fuels for the domestic and export markets are produced.

Seen passing through Pantyffordd in the Dulais Valley on 16 July 2019, No. 66023 starts the long journey to Immingham (6E09) with loaded coal from Onllwyn.

On 29 March 2019, No. 66023 pushes back four flatbed wagons at Onllwyn, each containing containers to be loaded with coal for the onwards journey to Mossend via Margam.

No. 66023 makes its way on 29 March 2019 to the washery at Onllwyn on the 6B28 from Swansea Burrows with anthracite brought in from Gwaun-Cae-Gurwen. The coal will be washed and blended at the plant for despatch to Immingham or Mossend.

On 29 March 2019, No. 66023, working the 6O20 Onllwyn to Swansea Burrows, is seen passing through Seven Sisters hauling a mixed consist of six empty MEA wagons at the rear of the train, behind eight loaded coal containers. The coal containers will go to Margam for the onward journey to Mossend. The MEA wagons will stay at Swansea Burrows Sidings. These are the last containerised coal shipments by rail in the UK.

Having just handed over the token, No. 66023 makes its way to Onllwyn on 29 March 2019 with a mixed consist on the 6B28 from Swansea Burrows. The train consists of six loaded MEA wagons containing anthracite brought in from Gwaun-Cae-Gurwen and Celtic Energy household coal containers.

No. 60015 slowly makes its way along the former Neath & Brecon line to the wash plant at Onllwyn with MEA wagons, loaded with anthracite from the Gwaen-Cae-Gurwen, in tow. The locomotive is on the second of two workings conveying coal from Swansea Burrows (6O08) on 5 September 2018.

On rare occasions, Class 60 locomotives from Margam were used to haul coal up from Swansea Burrows to Onllwyn in two journeys, each with a rake of 20 MEA box wagons. Here, No. 60015 can be seen on 5 September 2018 pulling forward a rake of empty box wagons, unloaded from its first trip up from Swansea Burrows. The train will continue for several hundred metres and then push back to couple with the MEA wagons brought from the second journey, which are being unloaded in the background. In a short while, the train will make the return journey to Swansea Burrows with all 40 unloaded MEA wagons in tow (6O10).

On 31 August 2018, DBC 'Red Shed' No. 66021, with empty MEA wagons in tow, passes through the little photographed Neath & Brecon Junction, destined for Swansea Burrows from Onllwyn (6O10). The line on the right heads to Cwmgwrach but has not seen a train since 2013, when the Unity mine went into administration.

No. 66085, working the Immingham to Onllwyn coal empties (4V32) on 17 May 2018, crosses the Neath River Swing Bridge near Skewen on the Swansea District line. The train will continue to Swansea Burrows where the locomotive will run round its trains for the onward journey to Onllwyn. The rusting 1903-built lightship, seen to the left, was in service on D-Day as a marker for the safe passage through a minefield for the invasion landing craft.

On 26 January 2018, an early morning start sees anthracite coal being unloaded from a rake of 20 MEA wagons at Onllwyn's wash plant. The coal, from East Pit in the Amman Valley, was hauled by train the previous day to Swansea Burrows for the onward journey to Onllwyn. At Swansea Burrows, the train of 40 loaded box wagons was split into two halves, with this being the first half brought in by No. 66115, working the 6O07. The locomotive will soon return light engine to the sidings at Swansea Burrows to collect the second load. The anthracite will be processed at the wash plant for despatch to Immingham to the Coal Products Group briquette factory.

No. 66023 passes through Pantyffordd, on 29 March 2019, with a mixed consist of six MEA wagons loaded with anthracite from Gwaun-Cae-Gurwen for processing at Onllwyn. There are also eight empty container wagons in tow, which will be loaded with coal for dispatch to Mossend via Margam.

On the cold morning of 10 January 2018, the mist gradually begins to lift as No. 66103 unloads the last of the coal from its MEA wagons at the Onllwyn washery. The coal is from Celtic Energy's East Pit opencast coal mine near Gwaun-Cae-Gurwen in the Amman Valley.

The Onllwyn washery pictured at the break of dawn of 26 January 2018. In this scene, coal is unloaded from MEA wagons brought in by DBS locomotive No. 66115. This is the first of two train loads of coal brought in from Swansea Burrows for processing. The coal has come in from the Amman Valley by rail on the Brynamman branch to Pantyffynnon and then via the Heart of Wales and Swansea District lines to Swansea Burrows for the onward journey to Onllwyn.

The train driver reads the newspaper in-between occasional episodes of moving forward as his train is loaded with Wales' 'finest black stuff' at the Onllwyn washery on 29 March 2019. This is the 6E09 to Immingham with No. 66004 in charge.

This scene shows the open, somewhat desolate, countryside at Onllwyn as DBC 'Shed' No. 66004 loads up with coal at the wash plant for the onward journey to Immingham (6E09) on 7 September 2017.

Rosebay willowherb adds a bit of colour to this scene as No. 66004 loads up with coal for the long haul to Immingham on the 6E09 on 7 September 2017.

On 31 August 2017, No. 66221 departs Onllwyn for the long haul to Immingham (6E09) with 21 loaded HTA wagons containing around 1,500 tonnes of coal.

No. 66161 with a rake of empty HTA wagons approaches the Onllwyn washery from Margam with empty coal hoppers (4O32) on 11 April 2017. The train will load with coal for the return journey to Immingham (6E09).

No. 66161 departs Onllwyn on the 6E09 to Immingham with 21 loaded HTA hoppers on 11 April 2017.

DBC 'Red Shed' No. 66161 rounds the curve as it departs Onllwyn washery hauling around 1,500 tonnes of anthracite, working the 6E09 to Immingham in Lincolnshire on 11 April 2017. The green strip of land left of the bridge is a remnant of the former Neath & Brecon line. The line survived as a freight-only branch to serve Craig-y-Nos Hobbs limestone quarry until its closure in 1977.

No. 66023 approaches Onllwyn, on 29 March 2019, with a mixed consist comprising six MEA wagons with coal brought in from Gwaun-Cae-Gurwen and eight coal containers on flatbed wagons. This is the 6B28 from Swansea Burrows. The coal will be processed at the plant, whilst the containers will be loaded for onward dispatch to Mossend via Margam.

On a rather dull, dismal wet day (4 April 2017) in the Dulais Valley, No. 66093 approaches the washery at Onllwyn on the 4O32 Margam to Onllwyn coal empties. The train will load with coal and return to Immingham on the 6E09.

On 4 April 2017, No. 66093 runs round its train at the Onllwyn washery to head the train on the right. The train will load up with around 1,500 tonnes of coal for the journey to Immingham (6E09).

No. 66161 is seen on 11 April 2017 on the approach to Onllwyn washery hauling empty HTA coal hoppers from Margam (4V32). The train will load up with coal (anthracite) for the return journey to Immingham (6E09). On the right-hand side, a mine dump truck makes its way to Nant Helen open-pit coal mine to dispose of waste from the wash plant. The hill in the background is a slag heap made from the overburden mined from the pit.

No. 60015 arrives at Onllwyn, on 5 September 2018, with the second load of coal brought from Swansea Burrows on the 6O08. The coal is from Celtic Energy's East Pit at the Amman Valley, loaded up at Gwaun-Cae-Gurwen.

A view from Onllwyn looking towards the west on 5 September 2018. Nowadays, it is it quite unusual to see coal tips, though they once dominated the landscape in South Wales. There are some exceptions though, in the few active coal-mining areas that still exist in South Wales, and this is one of them. The gentle hill beyond the power lines is a coal tip from the still-active Nant Helen opencast mine, run by Celtic Energy. The road on the right leads to the pit and is used to despatch coal by dump trucks to the Onllwyn plant for processing. The train in the scene consists of 40 empty MEA box wagons headed by No. 60015 and is destined for the sidings at Swansea Burrows (6O10).

The last few wagons of coal are unloaded from a train of MEA wagons at the Onllwyn processing facility in the Dulais Valley, South Wales, on 26 January 2018. The pay dirt is top quality anthracite brought by rail from the Gwaen-Cae-Gurwen loading bay near East Pit opencast mine in the Amman Valley. No. 66115 will head the train of empties back to Swansea Burrows (6O10). The coal will be processed at the wash plant for subsequent despatch in HTA wagons to the briquette factory at the Port of Immingham.

No. 60015 passes through Pantyffordd, on 5 September 2018, on the approach to Onllwyn along the former Neath & Brecon line, hauling anthracite from the Gwaen-Cae-Gurwen opencast coal mine (operated by Celtic Energy). The coal will be washed and graded at the Onllwyn processing plant. The train has come up from Swansea Burrows on the 6O08, the second of two workings that day.

On 26 January 2018, a mine dump track crosses the line at Onllwyn with plant waste for disposal at the nearby Nant Helen opencast coal mine. Meanwhile, No. 66155 delivers a second load of coal, brought from Gwaen-Cae-Gurwen on the 6O08 from Swansea Burrows, for processing at the washery. Though Nant Helen mine was mothballed in 2016 due to loss of demand for coal, approval for the mine to extend operations from its original deadline of December 2018 to December 2021 was granted by Powys County Council in March 2019. This has meant a temporary reprieve for the mining/processing operations at Onllwyn.

No. 66070 departs the washery at Onllwyn on 5 April 2016 with 21 HTA hoppers loaded with coal destined for Immingham on the 6E09.

No. 66115, working the 6O10, departs Onllwyn washery on 26 January 2018 with 40 empty MEA wagons for Swansea Burrows.

On 21 February 2014, No. 66221, with 21 loaded HTA's in tow, forming the 6O32 Onllwyn–Aberthaw power station, is seen passing through Neath & Brecon Junction. For many years, Onllwyn supplied Aberthaw power station, one of its principal customers, with coal; 9 September 2016 saw the last ever coal train to Aberthaw. The disused line to Cwmgwrach can be seen on the right.

No. 66008 heads south past Neath & Brecon Junction towards Swansea Burrows on the 6O32 Onllwyn–Aberthaw power station on 29 January 2014.

The driver of No. 66008 hands over the token to the signalman at Neath & Brecon Junction signal box on the 6O32 Onllwyn–Aberthaw power station on 29 January 2014.

Having just handed over the token to the signalman at Neath & Brecon Junction signal box, No. 66121, with 21 loaded HTA's in tow, heads south towards Swansea Burrows on the 6O32 Onllwyn–Aberthaw power station on 16 July 2014.

On 5 September 2018, No. 60015 passes through Pantyffordd, working the 6O08 from Swansea Burrows to Onllwyn with 20 MEAs in tow, loaded with coal from Gwaun-Cae-Gurwen for processing at the wash plant.

On 29 March 2019, MEA wagons, brought in from Swansea Burrows by No. 66023, are unloaded at the Onllwyn washery. Interestingly, the consist also comprises eight coal containers brought in from Margam. These will be loaded with coal for despatch to Mossend in Scotland.

On 7 September 2017, No. 66004 passes through Pantyffordd on the approach to Onllwyn Washery, working coal empties from Margam (4O32). The train will load up with coal for the return journey to Immingham on the 6E09.

On 5 September 2018, No. 60015, a Class 60, departs Onllwyn for Swansea Burrows with a rake of 40 empty MEA wagons, having discharged its load of coal at the wash plant.

On a beautiful summer day (16 July 2019), No. 66023 passes through Pantyffordd as it nears its destination of the washery at Onllwyn with a rake of 21 empty HTA wagons in tow from Margam (4O32) to be loaded with coal for the return journey to Immingham on the 6E09.

On 28 January 2020, having run round its train at Swansea Burrows, No. 66098 heads north near the village of Jersey Marine, about 5km east of Swansea, with 21 loaded HTA wagons destined for the long haul to Immingham via Margam on the 6E09.

Chapter 6
Gwaun-Cae-Gurwen

Gwaun-Cae-Gurwen (GCG) is a village located in the borough of Neath Port Talbot in South-West Wales. The line to GCG is a single track that branches off from the still semaphored Heart of Wales section of line at Pantyffynnon, 0.8km south-west of Ammanford in Carmarthenshire. It runs alongside the River Amman for approximately 11km on the former Brynamman Valley branch (Llanelli Railway), which ceased services to passengers on 18 August 1958, ahead of the Beeching Act. The line stayed in use to haul coal from collieries, exploiting the rich anthracite bearing seams in the Amman Valley, at the western extremity of the South Wales coalfield. It continued to operate until 1991 after the closure of Abernant Colliery in 1988, where coal from the Betws Drift Mine at Ammanford was processed using the colliery's wash plant facility.

This was not the end of the story though. Mining at Celtic Energy's East Pit opencast coal site, near the village of Brynamman, resumed operations in 2005. The 400ha site, which has been mined since the 1980s, had 2.1 million tonnes of mine reserves comprising high-quality anthracite coal in 12 individual seams lying up to 150 metres below the surface. Production was stepped up in 2008 with the introduction of new mining machinery, and it was reported that the site extracted 5,000 to 7,000 tonnes of coal per week.

In order to facilitate the transport of the coal and offset CO_2 emissions, the branch line to GCG was reopened to coal traffic in 2009. Freight trains hauling up to 1,300 tonnes of coal removed the need for at least 40 lorry loads of coal, using local roads from the coal mine, and resulted in four times lower CO_2 emissions than road haulage. The re-opening of the line enabled Celtic Energy to transport 50 per cent of its coal production via freight trains. This equated to around 300,000 tonnes or more of coal transported out of East Pit each year.

Although the line was never formally closed (it was not lifted), its reinstatement was not without its challenges. Network Rail started an awareness campaign by contacting local people to warn them of the dangers of trespassing on the line given it had been out of use for many years. Local residents and dog walkers were urged to observe safety rules at level crossings and to cross the line carefully. The line was upgraded at a cost of £1.2m.

DB Cargo was contracted by Celtic Energy to take the coal to Onllwyn in the Dulais Valley for processing. The first train to haul coal out of GCG was No. 66076 on 16 January 2009. It was the first to haul coal on the restored branch for 18 years. Class 66 locomotives were deployed to haul the coal in MEA wagons to Swansea Burrows for the onward journey to Onllwyn. Although Onllwyn is only about 14km east of GCG (as the crow flies), a somewhat convoluted route is required to get to Swansea Burrows. The route is along the Brynamman branch to Pantyffynnon, and then south along the Heart of Wales line to Hendy Junction, where a short curve links the Heart of Wales line to the Swansea District line. Swansea Burrows is reached via Jersey Marine North Junction, which branches off the Swansea District line.

Initially, the trains comprised 25 MEA wagons and ran once or twice a week, usually on a Wednesday/Friday basis. In later years, the trains increased in capacity to 40 MEA wagons (and sometimes 45 wagons) and ran on a Tuesday/Thursday basis. At Swansea Burrows, the wagons were split into two halves (20 wagons each) and taken to Onllwyn in two trains the following day (Wednesday/Friday) using a locomotive deployed from Margam.

For a brief time in 2016, coal trains also ran to Aberthaw power station in May/June and again in September/October, operated by Freightliner.

A token system, controlled by Pantyffynnon signal box, is used for the coal trains entering the branch to GCG. It is the only active box on the Heart of Wales line and controls the entire stretch of line from Llandeilo Junction, east of Llanelli, to Craven Arms in Shropshire. It uses a No-Signalman Token Remote signalling system developed in the 1980s to reduce running costs on lightly used single-track railways in rural areas. The box is a type GWR 5 which opened in 1892. It is fitted with a 49-lever frame and is a listed building.

The revamped branch saw little more than ten years of use due to the depletion of the coal reserves at East Pit, where the mining operation is now at an end. The final train to Swansea Burrows departed on 9 May 2019, worked by No. 66171.

After this, what coal remained was largely sent by truck, but some coal was also despatched by rail to Immingham in rakes of 20 HTA wagons. Workings were sporadic, run as and when they were required, every couple of months or so. The flow represented the very last of the coal trains on the former Brynamman branch line. The last train ran on 12 December 2019 with No. 66158 in front.

Pantyffynnon is well worth a visit, with its surviving signal box, extant semaphore signals and little station. The charming station, a grade II listed building, is believed to date back to 1857 and has been disused since the 1960s. It was recently renovated by Network Rail with support from Carmarthenshire's Heritage Team and significant grant funding from the Railway Heritage Trust. In 2017, the station was awarded the Railway Heritage Trust Award for the 'best restored structure' at the National Rail Heritage Awards.

The future of the Brynamman branch line remains bleak due the closure of East Pit. It is unlikely to see any more coal trains thus marking the end of a long history of coal mining and rail transportation in the anthracite-rich Amman Valley.

On 12 May 2014, No. 66096, with MEA box wagons, passes Pantyffynnon signal box on its journey to Gwaun-Cae-Gurwen on the 6G05 from Swansea Burrows. The train, loaded with coal, will return via Swansea Burrows to the wash processing plant facility at Onllwyn.

On 15 March 2018, DBC 'Red Shed' No. 66115 approaches Pantyffynnon signal box on the Heart of Wales line, working the 6G05 from Swansea Burrows to Gwaun-Cae-Gurwen to load with coal.

On 15 March 2018, the driver of No. 66115 collects the token at Pantyffynnon signal box for the journey to Gwaun-Cae-Gurwen from Swansea Burrows (6G05) to collect coal mined from East Pit in the Amman Valley.

Another view of 'Red Shed' No. 66115, pictured on 15 March 2018, looking north at Pantyffynnon station with MEA box wagons in tow for loading at Gwaun-Cae-Gurwen (6G05). The attractive station, a Grade II-listed building, dates back to 1857 but has been disused since the 1960s. The station building, which has been renovated recently, is a rare example of a 'Brunel chalet', which was a standard design by the famous railway engineer, Isambard Kingdom Brunel.

One of the last trains to use the branch line to Gwuan-Cae-Gurwen, No. 66024 is seen on 29 August 2019, with a rake of 20 empty HTA wagons in tow, on the 4V32 from Immingham passing the old Glanamman Station signal box, which is now used as an office by the Amman Railway Society. The old signal box is a Great Western Railway type 7b design, believed to date from 1901. The train will return loaded with coal a few hours later for the return journey to Immingham (6E09).

No. 66026 runs round its train at the Gwuan-Cae-Gurwen loading facility on 29 August 2019. It will take approximately 2.5 hours to load the train with coal ready for the return journey to Immingham on the 6E09.

Seen on 29 August 2019, No. 66026 has just arrived at Gwaun-Cae-Gurwen on the 4V32 from Immingham with empty HTA hoppers to load up with anthracite. This was one of the last trains to Gwaun-Cae-Gurwen.

Gwaun-Cae-Gurwen

On 21 June 2018, DBC 'Red Shed' No. 66082, with 40 MEA wagons filled with anthracite in tow, makes its way across the Pontarddulais Viaduct at Morlais Junction. The viaduct, built in red engineering brick by the GWR as part of the Swansea & District line, spans the River Loughor. It is some 200 metres long, rises over 20 metres above the river and has 11 semi-circular arches, each spanning some 30 metres. The train, working the 6O70 from Gwaun-Cae-Gurwen, is on its way to Swansea Burrows, where it will be split in half into two trains for the onward journey to Onllwyn the following day.

A shot of Pantyffynnon station, taken on 10 February 2014, prior to its renovation, with work in progress to restore the building to its former glory. The Heart of Wales line is seen on the left and the freight branch to Gwaun-Cae-Gurwen is on the right.

Chapter 7
Aberthaw Power Station

Situated on the coast, in the Vale of Glamorgan, between Barry and Rhoose, Aberthaw 'A' power station, owned by the Central Electricity Generating Board, officially opened in 1963 and was the most advanced in the world at that time. Aberthaw 'B' power station, today owned by German utilities company, Rheinisch-Westfälisches Elektrizitätswerk (RWE), was commissioned in 1971 and operated until its closure in December 2019.

Aberthaw 'A' operated until 1995 and was subsequently demolished in 1998, when the station's two chimneys were blown up. Aberthaw 'B' was saved from closure in June 2005 and was the last working coal-fired power station in Wales.

The site had three generators with a capacity of around 1,560 MW, enough to power the needs of 1.5 million households, or equivalent to the population of five cities the size of Cardiff. At its peak in 2013, the coal-fired power plant generated enough electricity to keep the lights on in 3 million homes every year. From 2006 to 2007, new steam turbines were fitted, allowing each unit to generate initially an extra 28–30 MW of power, subsequently upgraded to 520 MW. Up to 2 million tonnes of coal could be stored, which could either be stockpiled or conveyed directly into the mill feed hoppers.

Aberthaw 'B' burned approximately 5,000–6,000 tonnes of coal a day. It was designed specifically to burn bituminous (steam coal) and the semi-anthracitic low-volatile Welsh coal peculiar to the South Wales coalfield, providing a crucial lifeline for the coalfield. It was the only coal-fired power station in the UK designed to burn low- to mid-volatile coal. Two thirds of the coal burned was Welsh, the balance supplemented by imported coal. In later years, to comply with UK government directives, some power was also generated by burning woodchip biomass. According to the *New Statesman*, Aberthaw 'B' consumed some 2.3 million tonnes in 2013.

In order to make Aberthaw 'B' one of the most efficient coal-fired power stations in the UK, and to comply with European regulations, RWE invested in various technologies to reduce emissions. This included seawater flue gas desulphurisation technology, designed to remove up to 95 per cent of sulphur dioxide emissions by 2008. It also installed new technology for carbon capture. Since 2012, the power station's CO_2 emissions were reduced by 60 million tonnes per year. In addition, there was the installation of low-nitrogen oxide technology to reduce emissions by over 90 per cent.

However, by 2015, the power station was still in breach of EU regulations. The European Court of Justice ruled that emissions of toxic nitrogen oxides (NO_x) from burning Welsh coal at Aberthaw were too high, though RWE maintained it was fully compliant with the Industrial Emissions Directive (IED) to control emissions. The IED is the main EU instrument regulating pollutant emissions from industrial installations, adopted in 2014.

The European commission brought the case to court in 2015, on the grounds that not enough had been done to address excessive emissions of NO_x from the plant since concerns were first raised in 2013. The Department for Environment, Food and Rural Affairs (Defra) and the Welsh government, however, maintained they could use an exemption in the law for Aberthaw 'B' as it was burning coal with a volatile content of less than ten per cent. They argued the exemption was specifically designed for the power station as it used locally mined bituminous and anthracitic coal from the South Wales coalfield. This was economically important to the local area but more difficult to burn. They added that in order to comply with the IED, additional works would be undertaken, including boiler improvements to achieve lower NO_x emissions.

The court was not convinced though. It pointed out that the annual average volatile content of Aberthaw coal was in excess of the ten per cent limit. It also concluded there was no official evidence that the exemption was designed with Aberthaw 'B' in mind.

Therefore, to obviate toxic emissions, the decision was made to stop using Welsh coal to generate power from 2017. Instead, high-volatile imported coal (mainly from Russia), which had fewer nitrogen and sulphur emissions, was brought in from Avonmouth to power the plant. RWE also made the decision to generate power only at peak times, mainly as and when required during the winter months.

As a result of difficult market conditions, RWE announced that Aberthaw 'B' would shut down in August 2019. On 13 December 2019, the plant finally closed, ending almost half a century of power generation at Aberthaw 'B'. It is scheduled to be decommissioned in March 2020.

Rail Traffic

By the early 1960s, coal started arriving at the power station from a newly built 2.8km (1¾ miles) long branch that ran from East Aberthaw station on the Vale of Glamorgan line. The line forms a pear-shaped loop to the power station for the trains to discharge their coal and return to the reception sidings, which can hold two trains at a time. The section of line used to have a splendid array of semaphore signals controlled by Aberthaw signal box, until it closed on 9 March 2013, after 116 years of service. It is now a grade II listed building, located on the platform of the old Aberthaw station.

Aberthaw Freight Diagrams – August 2014			
Monday–Saturday	Origin	Time	Headcode
MWFO	New Cumnock	17:22/18:08	6V22
MWFO	Avonmouth	09:24	6B68
	Cwmbargoed	14:00	6C93
MWFO	Onllwyn	12:16	6O32
	Onllwyn	20:10	6O11
	Tower	10:54	6C45
MX	Avonmouth	05:50	6B66
SX	Tower	18:17	6C47
MX	Avonmouth	07:10	6B73
SX	Newport Docks	08:30	6F57
SX	Avonmouth	08:07	6B70

MWFO – Mondays, Wednesdays, Fridays; MX – Mondays excepted, i.e. ran Tuesdays to Fridays; SX – Saturdays excepted, i.e. ran Mondays to Fridays

For around 30 years, until the early 1990s, Aberthaw was mainly supplied with coal from Welsh collieries in trains hauled by the stalwart Class 37 locomotives, in rakes of 28 HAA hoppers (or more), each with the capacity to hold 29 tonnes of coal. In June 1979, there were 14 loaded trains recorded per weekday, which by the start of the next decade increased to 20 loaded trains bringing coal down the valleys each day. Assuming an average rake of 28 hoppers, each train would haul around 812 tonnes of coal at that time. In contrast, the modern trains hauled by Class 66 locomotives, in rakes of 21 HTA hoppers (76-tonne capacity), carry almost twice the amount of coal, equating to 1,596 tonnes per train.

After the privatisation of the railways, EWS (later DB Cargo) pretty much had the monopoly as the FOC to Aberthaw for many years, using Class 66 locomotives (introduced in 2008) to haul the coal in rakes of purpose-built HTA bogie coal hoppers. Freightliner replaced DB Cargo as the FOC in April 2016. Prior to this, Colas operated on an intermittent basis in August 2014 and again from December 2014 to February 2015, via Gloucester NY.

Two thirds of the coal supplied to Aberthaw was from opencast operations in South Wales, sourced from Tower Colliery (later replaced by the open pit Tower Regeneration Project), Cwmbargoed and Onllwyn. Coal flows from Cwmgwrach ceased in 2013, when the Unity mine went into liquidation. Coal was also imported from Newport and Avonmouth Docks. The number of diagrams varied according to consumption requirements, but as many as ten trains per day were reported in August 2014 during a busy period. Coal was even brought in from as far away as New Cumnock in East Ayrshire, Scotland. This was the longest distance rail-borne coal flow in the UK at that time, involving a journey of some 585km (364 miles) and requiring three crew changes. It was testament to the importance of keeping the power station fully supplied at that time.

There were dramatic changes to the operation in 2017 in light of RWE's decision to use imported coal to replace Welsh coal, ending more than half a century's use of coal supplied from the South Wales coalfield. The final trains to take Welsh coal to Aberthaw ran from Onllwyn on the 9 September 2016, from Tower on 24 February 2017 and from Cwmbargoed on 15 March 2017.

However, this was not quite the end of the story. Colas re-commenced the movement of imported coal (mainly from Russia) from Avonmouth on 1 August 2017, after a four-month hiatus of rail-borne traffic. Class 70 locomotives were used on the flow with up to two trains a day. Nonetheless, the contract was short lived, ending on 28 February 2018.

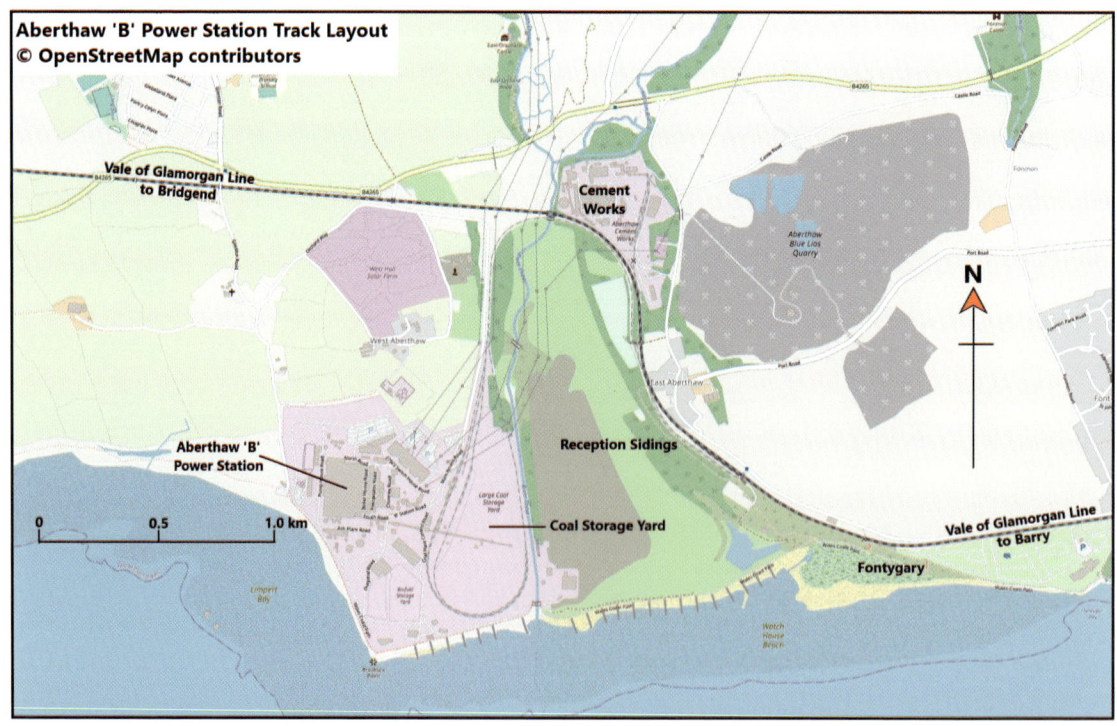

Whilst this was the end of the regular coal flow to Aberthaw, six more trains from Cwmbargoed, operated by Freightliner in late September/early October 2018, made the journey to the power station bringing in coal for blending purposes. Ten months later, on 9 August 2019, the last ever train ran from Portbury via East Usk Junction in Newport, headed by Freightliner Class 66/5 No. 66504. Bizarrely, there was just the one train after months of inactivity.

Access to the power station is via the scenic Vale of Glamorgan line (VOG), which runs from Cardiff to Bridgend. The line opened in 1895 for the transportation of coal to the port of Barry, limestone from local quarries and cement from Aberthaw. The line also provided passenger services from Barry to Bridgend. The eastern section was quite busy due to the large RAF base at St Athan, but the western section carried little traffic and fell victim to the Beeching Act in 1964. The line stayed open for freight traffic, however, and enjoyed a reprieve with the opening of Aberthaw power station in 1963, which ensured its survival. It is strategically important as a relief line for freight and passenger traffic, particularly when engineering work takes place on the main line between Bridgend and Cardiff, usually during the early hours and on Sundays.

There were several campaigns to have the line re-opened for passengers, but nothing happened for many years. However, with increasing traffic to Cardiff airport, the local government transport consortium, SWIFT, identified the potential for reopening the line in 1999. The Vale of Glamorgan and Bridgend Councils promoted this to the Welsh Assembly and the line was finally reopened to passenger traffic again on Sunday 12 June 2005, the first day of that year's summer timetable. This provided a major boost to the VOG line, ensuring its survival.

The line follows a particularly scenic route between Barry and Aberthaw. The impressive Porthkerry Viaduct, which extends across Porthkerry valley, spans some 16 arches, each between 45 and 50 feet (15 metres) in width, and rises to a height of 110 feet (33 metres). Construction began in 1896 but problems caused by subsidence in 1896 resulted in the temporary closure of the viaduct. It finally opened to traffic in 1900. From Barry, the line climbs at a gradient of 1 in 81 to the Porthkerry Viaduct and then descends at 1 in 165 to 1 in 200 beyond Aberthaw. The steep gradient to Aberthaw is hard work for the coal trains, perhaps not surprising as they have over 2,000 tonnes to move.

On the approach to Aberthaw, at Fontygarry, the railway follows a splendid stretch of unique coastline, with its salt lagoons, sand dunes, rock pools, sea grass and fabulous views of the Somerset coast and Bristol Channel. Fontygarry was voted 18 of the top 25 best railway locations to photograph in the UK by the *Rail Express* magazine in May 2014 (edition No. 216), so it is well worth a visit, with or without the trains!

After the decision to stop burning Welsh coal at Aberthaw 'B' power station, imported coal (mainly from Russia) was brought in from Avonmouth from 1 August 2017. The contract was short-lived though, ending on 28 February 2018 when coal rail traffic ceased on a regular basis. In this 9 August 2017 scene, No. 70816 is seen arriving at the power station loaded with coal from Avonmouth (6B22).

On 26 September 2017, No. 70803 negotiates the 180-degree curve with imported coal from Avonmouth (6B22) for discharge at Aberthaw power station.

The final Freightliner train to Aberthaw left Cwmbargoed on 15 March 2017. Whilst this was the end of the regular coal flow to Aberthaw, six more trains, operated by Freightliner in late September/early October 2018, made the journey with coal brought in for blending purposes. In this scene, on a dull afternoon on the 2 October 2018, 66/5 No. 66516 arrives at the power station with around 1,500 tonnes of Welsh coal from Cwmbargoed (6C93). It was one of the last trains to bring coal to the power station.

Amongst the pylons and overhead power lines, No. 70816 discharges its load of coal, brought in from Avonmouth (6B22), at Aberthaw 'B' power station in the Vale of Glamorgan on 9 August 2017.

It's rare to see a Class 60 locomotive at Aberthaw, but on 3 October 2017, No. 60056 enters Aberthaw power station, working the 6V94 Lindsey oil refinery loaded oil tanks train. The power station used a small quantity of fuel oil for the start-up and stabilisation of the boiler combustion process.

No. 70803 gently edges forward as it discharges its trainload of imported coal from Aberthaw (6B22) on 3 October 2017. Note the stockpiled coal in the background, which has built up considerably after just a few weeks of receiving coal from Avonmouth for use in power generation during the winter months. Meanwhile, unusually, there is another locomotive at the power station, Colas No. 60056, with loaded oil tanks brought in from Lindsey oil refinery (6V94). The tanks are being discharged for the start-up and stabilisation of the boiler combustion process.

No. 70816 approaches the reception sidings at Aberthaw on 8 August 2017, with a train from Avonmouth (6B22). The last train to supply Welsh coal to Aberthaw came in from Cwmbargoed on 15 March 2017. The power station then relied on imported coal to replenish the coal stock. A new contract with Colas to supply the coal commenced on 1 August 2017. Up to two trains a day, using Class 70 locomotives, brought the coal in from Avonmouth, but the contract was short-lived, ending on 28 February 2018.

The light fades on 10 November 2014 as No. 66107 brings in coal to Aberthaw from Onllwyn on the 6O32. The smoking stack at Aberthaw 'B', on the left, was problematic for the power station due to toxic NOx emissions.

Aberthaw reception sidings on 20 January 2014, with DB Cargo's No. 66135 stabled on the right with coal empties for its return journey to Onllwyn (4O11). Meanwhile, No. 66144 has just arrived at the sidings with coal empties, having discharged its coal at the power station, and will soon depart to Cwmbargoed, working the 4C95.

The scene at Aberthaw reception sidings on 6 February 2020. The rusty tracks on the left have not seen regular use since the coal flow from Avonmouth stopped on 28 February 2018. Coal trains still use the Vale of Glamorgan line on a regular basis on their way to Port Talbot and here No. 66053 passes the redundant sidings with coal empties from Margam destined for Cwmbargoed (4C93).

In busier times, MGR trains make their way to and from Aberthaw 'B' power station at the start of the loop branch to the power station on 13 November 2013. Note the Vale of Glamorgan line on the left.

No 66116 is seen at the reception sidings at Aberthaw with coal empties destined for Avonmouth (4C96) on 10 November 2014.

Class 60 traction is rare on the Vale of Glamorgan line. However, here No. 60039, working the Aberthaw power station to Lindsey oil refinery empty oil tanks (6E94), can be seen waiting for its path to proceed at the reception sidings on 10 November 2014. Aberthaw signal box, which is a listed building and was decommissioned in early 2013, can be seen in the background.

EWS-liveried 'Tug' No. 60065, *Spirit of Jaguar*, rounds the curve on the approach to Barry Station, working the 6Z77 Cwmbargoed to Hope (Earles Sidings) loaded coal train via Margam, late in the evening of 27 June 2014. The signalman takes the opportunity to photograph this rare event – it had been many years since a Class 60 locomotive worked a coal train in South Wales. This is a poignant shot as this was the last working day for Barry signal box after 117 years of operational service. At the time, the box controlled the signalling to Cadoxton, Barry Island and along part of the VOG line to Aberthaw. An hour after this picture was taken, the signal box closed.

No. 66102 passes Barry signal box on the 6C93 Cwmbargoed–Aberthaw power station loaded coal service. The signal box was closed four days before this picture was taken. The structure was a Barry Railway Type II signal box, built for the Barry Railway Company, dating back to 1897. It was demolished in March 2015.

No. 66094 heads west towards Aberthaw power station, having just passed through Barry Station, working the 6C45 Tower Colliery loaded coal service on 14 June 2014. The semaphore signal was one of three remaining signals controlled by Barry signal box on the branch line to Barry Island at that time. They have since been replaced with LED lights.

No. 70814 passes through the railway cutting between Barry Docks and Cadoxton station, working the 4C21 Aberthaw power station–Avonmouth coal empties on 18 December 2017. Below, is the single-track line to Barry Docks, which passes the remnants of an old semaphore signal gantry.

Patchy sunlight illuminates the Bristol Channel on an otherwise grey, dull December afternoon (10 December 2013), whilst No. 66081 makes its way to Aberthaw power station along the Vale of Glamorgan line on the 6C45 loaded coal from Tower.

On a beautiful summer morning, No. 66053 rounds the curve on the approach to Cadoxton station in Barry, working the 4C93 Margam to Cwmbargoed coal empties on 3 June 2019.

DB Cargo 'Shed' No. 66027, powering the 6C45 Tower–Aberthaw power station loaded coal service, is seen in the railway cutting between Barry Docks and Cadoxton station on 24 February 2014. The line on the right sweeps off to form a 180-degree curve into Barry Docks.

With a view of Aberthaw 'B' power station in the background, No. 66200 passes along the scenic coast at Fontygary in the Vale of Glamorgan, working the 4V01 Hope (Earles Sidings) to Cwmbargoed coal empties on 18 October 2019. This is a once (sometimes twice) weekly working from Hope cement works in Derbyshire.

Class 66 locomotives have been used exclusively to haul the coal out of Cwmbargoed, but there have been one or two exceptions in recent years when Class 60 locomotives have made a rare appearance. On the 27 June 2014, No. 60065, *Spirit of Jaguar*, was the first Class 60 to arrive at Cwmbargoed since 1998. The locomotive is seen hauling coal empties from Hope in Derbyshire, working the 4V01, passing along the scenic coastline at Fontygarry in the Vale of Glamorgan en route to Cwmbargoed to load up with coal.

Taken just after sunset on 12 November 2016, orange-yellow skies appear over the Bristol Channel, while the last light illuminates the rock pools along the Glamorganshire coastline at Fontygary. The last train heads out east from Aberthaw power station along the Vale of Glamorgan line towards Barry, with a rake of empty coal wagons destined for Stoke Gifford on the 4C96 and 66/5 No. 66553 in charge.

Red sky at night, sailors delight! On 30 August 2016, 66/5 No. 66546 slowly makes it way along the coastline at Fontygary with a rake of empty HAA wagons from Aberthaw power station (seen in the background) destined for Stoke Gifford on the 4C96.

No. 66099, seen from Fontygary caravan park in the Vale of Glamorgan, heads the 4C61 Aberthaw–Avonmouth coal empties on 13 November 2013. Interestingly, there is a headboard displaying the headcode of the train in the cab, a rare sight.

No. 66081 nears the end of its 6 December 2013 journey, as it heads along the Vale of Glamorgan coastline at Fontygary to Aberthaw 'B' power station with a loaded coal train from Tower on the 6C45.

On a beautiful summer evening (21 July 2014), DB Cargo 'Shed' No. 66011 passes through St Athan in the Vale of Glamorgan, with a train headed west towards Bridgend for Margam. This is the 6M77 Cwmbargoed to Hope (Earles Sidings) loaded coal train. Aberthaw power station is seen on the right in the background.

As the winter sun rises over Porthkerry Viaduct, an unidentified Class 66 locomotive heads a coal train destined for Aberthaw power station on the 6B66 from Avonmouth on 16 February 2016.

On a cold, crisp, frosty morning (17 February 2015), with the sun rising above the horizon, No. 66182 nears the end of its journey as it crosses Porthkerry Viaduct towards Aberthaw 'B' power station with loaded coal from Avonmouth (6B66).

On a balmy summer evening (25 July 2018), the late sun catches the arches on Porthkerry Viaduct in the Vale of Glamorgan as No. 66069, in charge of the Cwmbargoed–Scunthorpe B.S.C. loaded coal train (6E09), slowly makes its way across the impressive structure.

With the first hints of autumn, No. 66093 is viewed from Porthkerry Park, on 11 September 2014, as it slowly crosses the arches towards Aberthaw with a coal train from Avonmouth.

On the clear afternoon of 29 April 2015, with views of Steep Holm in the Bristol Channel and Brean Down, a promontory off the coast of Somerset, No. 66030 crosses Porthkerry Viaduct with a train laden with coal destined for Aberthaw power station from Cwmbargoed on the 6C93.

On a beautiful February afternoon (15 February 2019), 66/5 No. 66547 slowly crosses Porthkerry Viaduct in the Vale of Glamorgan, working the 6C94 Cwmbargoed to Port Talbot loaded coal. The line climbs at 1 in 81 from Barry to Porthkerry Viaduct and the steep gradient to Aberthaw is hard work for the coal trains, perhaps not surprising as they have over 2,000 tonnes to move.

There are some fine views of the Bristol Channel on a clear day from Barry. On 23 July 2018, No. 66039 heads across Porthkerry Viaduct on a still clear evening, working the 6M77 Cwmbargoed–Hope (Earles Sidings) loaded coal train against the backdrop of the Bristol Channel and Somerset coast.

In late summer evening light and very calm conditions, No. 66091 crosses Porthkerry Viaduct towards Barry, on the 4C96 Aberthaw–Cwmbargoed coal empties, at 20:28 on 7 July 2015.

Viewed from the south at Porthkerry Park near Barry on 23 February 2015, No. 66034 is seen crossing Porthkerry Viaduct on the 4V01 service, Hope–Cwmbargoed, to load with coal for the return journey to Hope (Earles Sidings in Derbyshire) via Margam.

No.66199 heads west towards Aberthaw power station working the 6C45 Tower–Aberthaw loaded coal service on 15 April 2014.

Against the backdrop of the mill-pond-like Bristol Channel, DB Cargo liveried 'Red Shed' No. 66014 hauls coal across Porthkerry Viaduct to Port Talbot Grange Sidings on the 6C94 from Cwmbargoed on 24 May 2017.

At 07:17 on 29 February 2018, winter sunrise at Porthkerry sees 66/5 No. 66504 crossing the viaduct, working the 4C91 East Usk Junction–Margam coal empties.

At 04:37 on 17 July 2018, diversions along the Vale of Glamorgan Line see 66/5 No. 66524, heading around 1,500 tonnes of coal destined for Port Talbot Grange from York Yard South (6V40), crossing Porthkerry Viaduct near Barry.

On 8 September 2014, two Class 60 locomotives made the journey to Cwmbargoed from Earles Sidings in Deryshire; this was the second such visit. Seen in fading light on the VOG line at Cogan Junction, No. 60001 hauls 18 loaded HTA wagons destined for Hope cement works in Derbyshire (via Margam) on the 6M77 from Cwmbargoed.

No. 66054 works the New Cumnock (in East Ayrshire, Scotland) to Aberthaw loaded coal train (6V22), seen coming off the relief loop at Cogan Junction, near Penarth, on 12 September 2013. This was longest distance coal flow in the UK at that time, involving a journey of some 585km (364 miles) and requiring three crew changes, testament to the importance of keeping the power station fully supplied.

Graffiti adorns the 140-metre-long wall at Seven Oaks Park in Cardiff (Grangetown). Street and graffiti artists come from all over Britain to decorate it with a mix of colours. Meanwhile, on 22 April 2015, No. 66170, with a load of coal from Onllwyn, joins the Vale of Glamorgan line from Penarth South Curve Junction for the onward journey to Aberthaw (6O31).

Whilst coal has long since left Barry Docks for export, coal trains still run through the town on a regular basis. In this scene, Freightliner Class 66/5 66953 passes through Barry Docks station, working the 6C45 Tower–Aberthaw loaded coal train on 27 April 2016.

A new building development near Llandough hospital afforded this view of Cardiff city on 30 April 2018. Freightliner 'Shed' 66/5 No. 66537 crosses the River Ely as it departs Cardiff (Grangetown) on the Vale of Glamorgan line, hauling coal destined for Port Talbot Grange on the 6C93 from Cwmbargoed. The building next to the flyover was once a Victorian pumping station but is now a Grade II listed building full of antique furniture and collectables.

The EWS-liveried locomotive and HTA wagons really stand out against the lush vegetation showing a hint of autumn colours along the banks of the River Ely, near Grangetown in Cardiff. No. 66012 hauls a rake of coal empties along the last section of the Vale of Glamorgan line before Cardiff, working the 4C96 from Aberthaw to Tower on 30 September 2014.

DB Cargo 'Red Shed' No. 66114 crosses the River Ely between Grangetown and Cogan Junction, near Cardiff, on the 6B70 Avonmouth–Aberthaw loaded coal on 19 January 2015.

On 14 January 2017, Freightliner 'Shed' 66/5 No.66599 passes through Eastbrook station, Dinas Powys, hauling a loaded coal train on the 6C45 Tower–Aberthaw.

On 21 January 2015, DB 'Shed' No. 66198 waits for the green signal to proceed to Newport ADJ with empty bogies, having unloaded its cargo of containers at Barry Docks (6B39), which can be seen in the background. The chemical tanker *Bergstraum* is berthed at No2 Dock quayside, discharging her chemical cargo to the Dow Corning plant, situated just out of sight to the left. Meanwhile, No. 66114, passes the ADJ train, working the 4C95 Aberthaw–Cwmbargoed coal empties.

Further reading from

As Europe's leading transport publisher, we produce a wide range of market-leading railway magazines.

Visit: shop.keypublishing.com for more details